AF595010

Algorithms in Decision Support Systems

Algorithms in Decision Support Systems

Editor

Vicente García-Díaz

MDPI • Basel • Beijing • Wuhan • Barcelona • Belgrade • Manchester • Tokyo • Cluj • Tianjin

Editor
Vicente García-Díaz
University of Oviedo
Spain

Editorial Office
MDPI
St. Alban-Anlage 66
4052 Basel, Switzerland

This is a reprint of articles from the Special Issue published online in the open access journal *Algorithms* (ISSN 1999-4893) (available at: https://www.mdpi.com/journal/algorithms/special_issues/decision_systems).

For citation purposes, cite each article independently as indicated on the article page online and as indicated below:

LastName, A.A.; LastName, B.B.; LastName, C.C. Article Title. *Journal Name* **Year**, *Volume Number*, Page Range.

ISBN 978-3-0365-0588-6 (Hbk)
ISBN 978-3-0365-0589-3 (PDF)

Contents

About the Editor

Vicente García-Díaz is an Associate Professor in the Department of Computer Science at the University of Oviedo. He is a Software Engineer, with a Ph.D. in Computer Science. He has a Master's in Occupational Risk Prevention and the qualification of University Expert in Blockchain Application Development. He is also part of the editorial and advisory board of several journals and has been an editor of several Special Issues in books and journals. He has supervised 100+ academic projects and published 100+ research papers in journals, conferences, and books. His teaching interests are primarily in the design and analysis of algorithms and the design of domain-specific languages. His current research interests include decision support systems, health informatics, and eLearning.

Preface to "Algorithms in Decision Support Systems"

Decision support systems (DSSs) are increasingly important information systems that help to make decisions related to unstructured and semi-unstructured decision problems that do not have a simple solution from a human point of view. They are currently used in different areas, such as medical diagnosis, catastrophe avoidance, agriculture, sustainable development, sales projections, inventory organization, production design, etc. The architecture of a common DSS is basically composed of three main components: (1) knowledge base; (2) user interface; and (3) model to infer the decisions. Such models may be based on multiple types of algorithms, such as neural networks, logistic regression, classification trees, fuzzy logic, etc. Although there are multiple works that try to optimize the operation of DSSs, researchers are still trying to optimize their performance by refining and proposing new algorithms that normally are adapted to the set of data available for a particular domain of knowledge. Thus, the aim of this book is to enhance the state of the art in this area significantly, as well as improving the performance of DSSs in specific domains.

The book is structured in such a way that the first works (articles 1-4) focus on some relevant knowledge domains and the last works (articles 5-8) deal with more general aspects on which researchers are actively working.

Thus, to start with, Fernando López-Martínez et al. [1] discuss a platform that is the new pillar for the Keralty Foundation to improve population health management, value-based care, and new upcoming challenges in healthcare. The benefits of using the new data platform include better healthcare outcomes, improvement of clinical operations, reduced costs of care, and the generation of accurate medical information. Several machine learning algorithms are used with standardized datasets to improve the effectiveness of public health interventions, improving diagnosis, and clinical decision support.

In the second work, Roanes-Lozano et al. [2] develop a prototype of a rule-based expert system aimed at an amateur competition player that is not accompanied by her coach to a championship. The player must answer a set of questions about how she is serving that day and her usual serving technique. Then, the system obtains a diagnosis using logic inference about the possible reasons. A certain knowledge of the tennis terminology and technique is required from the player, but that is something known at this level. The underlying logic is Boolean, and the inference engine is algebraic.

In the third work, Guimapi et al. [3] propose a platform that can improve insect pest management in the biological control context. It is an interactive platform for fitting data derived from experiments to mathematical expressions and carrying out spatial visualization. It uses experimental data as the input for model fitting, then applies the obtained model at the landscape level via a spatial temperature grid data to yield regional and continental maps. Different modules and functionalities of the tool are presented with the case study.

In the fourth work, Josyula et al. [4] present an evaluation framework for train rescheduling algorithms, which are very important whenever disturbances occur. They also present two train rescheduling algorithms: a heuristic and a mixed-integer linear programming-based exact algorithm. Finally, they conduct an experiment to compare the two multi-objective algorithms using a proposed framework. It is found that the heuristic algorithm is suitable for solving simpler disturbance scenarios since it is quick in producing decent solutions.

In the fifth work, Pendharkar [5] illustrates that, in addition to unbiased evaluations, the ensemble dimensionality reduction research in data envelopment analysis scores results in unique rankings that have high entropy. Under restrictive assumptions, it is also shown that the ensemble scores are normally distributed. Ensemble models do not require any new modifications to existing objective functions or constraints, and when ensemble scores are normally distributed, returns-to-scale hypothesis testing can be carried out using traditional parametric statistical techniques.

In the sixth work, Feng et al. [6] propose an unknown radar emitter identification method based on semi-supervised and transfer learning. Firstly, they construct a support vector machine model based on transfer learning, which can solve the problem that the training data and the testing data do not satisfy the same-distribution hypothesis. Then, they design a semi-supervised co-training algorithm, which can solve the problem that insufficient labeled data result in inadequate training of the classifier. Simulation experiments show that the proposed combination can effectively identify an unknown radar emitter.

In the second to last work, Koukoutsis et al. [7] list the more significant of the hazards and risks related to managing, updating, modifying, and upgrading the data and program cores of very large-scale DSSs. The authors also introduce a new general methodology for designing DSSs that are robust and circumvent these risks. The core of this new approach is the introduction of a meta-database, called teleological, on the base of which the management, updating, modification, reduction, growth, and upgrading of the system may be safely and efficiently achieved.

Lastly, Thanajiranthorn and Songram [8] suggest a new associate classification algorithm to directly discover a compact number of efficient rules for classification without the pruning process. A vertical data representation technique is implemented to avoid redundant rule generation and to reduce time spent in the mining process. The experimental results show that the proposed algorithm achieves in terms of accuracy a number of generated rule items, classifier building time, and memory consumption, especially when compared to the well-known algorithms.

I hope this book is to the liking of the reader and serves to deepen the knowledge of the exciting area of algorithms applied in decision support systems.

BIBLIOGRAPHY

[1] López-Martínez, F.; Núñez-Valdez, E.R.; García-Díaz, V.; Bursac, Z. A Case Study for a Big Data and Machine Learning Platform to Improve Medical Decision Support in Population Health Management. Algorithms 2020, 13, 102. https://doi.org/10.3390/a13040102

[2] Roanes-Lozano, E.; Casella, E.A.; Sánchez, F.; Hernando, A. Diagnosis in Tennis Serving Technique. Algorithms 2020, 13, 106. https://doi.org/10.3390/a13050106

[3] Guimapi, R.A.; Mohamed, S.A.; Biber-Freudenberger, L.; Mwangi, W.; Ekesi, S.; Borgemeister, C.; Tonnang, H.E.Z. Decision Support System for Fitting and Mapping Nonlinear Functions with Application to Insect Pest Management in the Biological Control Context. Algorithms 2020, 13, 104. https://doi.org/10.3390/a13040104

[4] Josyula, S.P.; Krasemann, J.T.; Lundberg, L. An Evaluation Framework and Algorithms for Train Rescheduling. Algorithms 2020, 13, 332. https://doi.org/10.3390/a13120332

[5] Pendharkar, P.C. A Comparison of Ensemble and Dimensionality Reduction DEA Models Based on Entropy Criterion. Algorithms 2020, 13, 232. https://doi.org/10.3390/a13090232

[6] Feng, Y.; Wang, G.; Liu, Z.; Feng, R.; Chen, X.; Tai, N. An Unknown Radar Emitter Identification Method Based on Semi-Supervised and Transfer Learning. Algorithms 2019, 12, 271. https://doi.org/10.3390/a12120271

[7] Koukoutsis, E.; Papaodysseus, C.; Tsavdaridis, G.; Karadimas, N.V.; Ballis, A.; Mamatsi, E.; Mamatsis, A.R. Design Limitations, Errors and Hazards in Creating Decision Support Platforms with Large- and Very Large-Scale Data and Program Cores. Algorithms 2020, 13, 341. https://doi.org/10.3390/a13120341

[8] Thanajiranthorn, C.; Songram, P. Efficient Rule Generation for Associative Classification. Algorithms 2020, 13, 299. https://doi.org/10.3390/a13110299

Vicente García-Díaz
Editor

Article

A Case Study for a Big Data and Machine Learning Platform to Improve Medical Decision Support in Population Health Management

Fernando López-Martínez [1,*,†,‡], Edward Rolando Núñez-Valdez [1,‡], Vicente García-Díaz [1,‡] and Zoran Bursac [2,‡]

1. Department of Computer Science, Oviedo University, 33003 Oviedo, Spain; nunezedward@uniovi.es (E.R.N.-V.); garciavicente@uniovi.es (V.G.-D.)
2. Department of Biostatistics, Florida International University, Miami, FL 33199, USA; zbursac@fiu.edu

* Correspondence: uo259897@uniovi.es; Tel.: +1-551-587-0112

† Current address: Sanitas Medical Center Corporate Offices, 8400 NW 33rd St, Doral, FL 33122, USA.

‡ These authors contributed equally to this work.

Received: 28 February 2020; Accepted: 21 April 2020; Published: 23 April 2020

Abstract: Big data and artificial intelligence are currently two of the most important and trending pieces for innovation and predictive analytics in healthcare, leading the digital healthcare transformation. Keralty organization is already working on developing an intelligent big data analytic platform based on machine learning and data integration principles. We discuss how this platform is the new pillar for the organization to improve population health management, value-based care, and new upcoming challenges in healthcare. The benefits of using this new data platform for community and population health include better healthcare outcomes, improvement of clinical operations, reducing costs of care, and generation of accurate medical information. Several machine learning algorithms implemented by the authors can use the large standardized datasets integrated into the platform to improve the effectiveness of public health interventions, improving diagnosis, and clinical decision support. The data integrated into the platform come from Electronic Health Records (EHR), Hospital Information Systems (HIS), Radiology Information Systems (RIS), and Laboratory Information Systems (LIS), as well as data generated by public health platforms, mobile data, social media, and clinical web portals. This massive volume of data is integrated using big data techniques for storage, retrieval, processing, and transformation. This paper presents the design of a digital health platform in a healthcare organization in Colombia to integrate operational, clinical, and business data repositories with advanced analytics to improve the decision-making process for population health management.

Keywords: decision support systems; population health management; big data; machine learning; deep learning; personalized patient care

---

1. Introduction

Colombia's health system is formed by the public sector and the private sector. The general social security system has two plans, contributory and subsidized. The contributory regime covers salaried workers, pensioners, and independent workers, with the subsidized plan covering anyone who cannot pay. Enrollment coverage increased from 96.6% in 2014 to 97.6% in 2015 [1].

The National Health Authority's primary purpose in Colombia is to improve the quality of healthcare and strengthening supervision, surveillance, and control of the health system. The 2015 Statutory Health Law No. 1751 places the responsibility for guaranteeing the right to health with the health system and recognizes health as a fundamental social right and makes it the state's responsibility to pursue an approach in health promotion and disease prevention [2].

The health sector in Colombia supports all initiatives for implementing new technologies to prevent cardiovascular diseases, disabilities, and high-cost hospitalization cases [3]. There is a remarkable need to improve the prediction of the risk of conditions for the population through the integration and unification of massive volumes of data and the implementation of effective advance analytic solutions to improve the decision-making process and population health management in Colombia's population [4].

Keralty organization is formed by a group of insurance and health services companies with a global presence, which together develops an integral health model, whose purpose is to produce health and well-being to people throughout their lives. The organization is committed to keeping its users healthy and autonomous, focusing on prevention, identification, and management of health risks, control, and care of disease and dependency [5]. The organization is a leader in Colombia by providing integrated health services and is recognized for their human, scientific, technical, and ethical approach [6].

This paper presents how we can obtain value from a large volume of heterogeneous data generated by different data sources in healthcare, and the architecture implemented. The development of proper advanced data analytics methods such as machine learning and big data analytics to perform meaningful real-time analysis on the data to predict clinical complications before it happens and to support the decision-making process are challenging but much needed to handle the complexity of the data-driven problems we are currently facing.

1.1. Related Work

Several initiatives in Europe, Asia, and North America aim to develop healthcare digital platforms with collaborative access tools to allow the exchange and sharing of information and knowledge wherever and whenever needed throughout the attention process. This type of frameworks and architectures will allow maximum quality and efficiency for patient's care, and to provide appropriate attention to the patient's condition and risks.

Castilla and Leon, for example, implemented a digitalization of health services as a tool to increase the efficiency of the services and increase the security in the attention to patient [7]. A healthcare cyber-physical system assisted by cloud and big data is being developed in the department of computer science at Pace University in New York [8]. This system consists of a data integration layer, a data management layer, and a data-analytics service layer to improve the functioning of the healthcare system. In France, a group of researchers implemented a wearable knowledge as a service platform to cleverly manage heterogeneous data coming from wearable devices to assist the physicians in supervising the patient health [9]. Another interesting work was presented at the International Conference on Computational Intelligence and Data Science (ICCIDS 2018). The authors proposed a hybrid four-layer healthcare model to improve disease diagnostic [10]. In India, a centralized architecture for an end to end integration of healthcare systems deployed in the cloud environment was developed using fog computing [11].

Medical organizations are investing more and more in developing a healthcare platform that integrates data, applications, business processes, and user interfaces to gain knowledge and useful insights for clinical decisions, drug recommendation systems, and better disease diagnoses. Some other examples of big data applications in healthcare can be found in healthcare monitoring, where data captured from wearable devices can assist providers in managing symptoms of patients online and adjust their prescriptions [12]. An analytical platform called "MedAware" has been developed to detect errors in medical prescriptions and clinical errors, reducing the hospital admission and readmission in real-time [13]. In the healthcare prediction field, a healthcare system called "Gemini (Generalizable Medical Information analysis and Integration system)" was developed to collect, process, and analyze large volumes of clinical data and apply machine learning algorithms for performing predictive analytics [14]. Other platforms have been implemented for genomics data analytics to generate predictions based on DNA molecular changes and mutations [15]. Another type of healthcare platform is related to the healthcare knowledge system, defined as the combination of clinical data and physician expertise to support clinical decision-making and diagnosis [16].

1.2. Why Big Data and Machine Learning?

Big data and machine learning are redefining healthcare goals for the future. Healthcare data are impacting the way disease research is performed, and the level of complexity in population health management is increasing as the traditional fee for service approach is transformed into the value-based care model [17,18].

Population health management is basically the aggregation of patient health data from multiple data sources, and the analysis and transformation into actionable insights to generate informed decisions to improve clinical and financial outcomes [19].

Big data technologies will allow us to bring large volumes of structured and unstructured data from disparate data sources into a data repository to be examined and analyzed. Machine learning models will assist in discovering insights from complex datasets with capabilities such as finding unseen patterns, making new predictions, and analyzing trends on health data. Machine learning is being used in a variety of clinical domains with the analysis of hundreds of clinical parameters resulting in effective and efficient models to improve the outcomes and quality of medical care models [20].

The implementation of this platform shows the enormous potential in using big data to individualize medical treatment, the opportunity for improving the lives of the patients, delivering better medical care, and reduced waste at an operational level [21]. Other chances for big data in healthcare for Keralty organization are:

- A physician would know before prescribing whether the patient is at high-risk to become dependent and different treatment plans can be selected based on this information.
- Psychosocial and clinical medical data could inform about the development of a chronic illness that can be properly diagnosed.
- The organization can use big data to understand how they are performing, the opportunities to improve clinical care, and their capacity to redesign care delivery to their patients.
- Using the platform's analytics component to improve the quality of care and patient experience at the lowest possible cost is core to the organization.
- Capturing streaming data and wearable data can provide to healthcare providers real-time insights about a patient's health that will allow them to improve their decision-making process for treatment and medication.
- Big data analysis can help the organization to deliver information that is evidence-based and can improve the efficiency, understanding, and implementation of the best practices associated with any disease.

In addition to the big data technologies used to build the platform, another essential component is the advanced analytic module of the platform. This module contains several machine learning algorithms to support clinical diagnosis. However, the organization should feel confident in these models and how they can be applied to specific use cases. These first models will alert providers to changes in high-risk conditions such as sepsis and hypertensive patients.

The main objective of this paper is to present the developed platform and its components to allow Keralty organization to derive better and more actionable insights from their data, i.e., to derive meaningful information from all these data in a way that allows them to improve care and lower costs needed for value-based reimbursement and business objectives while providing the highest quality care for population health management [22]. The goal is to be aligned with the triple aim framework developed by the Institute for Healthcare Improvement that describes an approach to optimizing healthcare system performance.The implementation of this platform intends to resolve several problems in health services to assist patients and their families in managing their health by providing better access to healthcare services [23].

2. Proposed Digital Health Platform

Keralty organization currently have several information systems such as Health Information Systems (HIS), Lab Information Systems (LIS), Radiology Information Systems (RIS), Enterprise Resource Planning (ERP), and Customer Relationship Management (CRM), among others, in their ambulatory care centers, hospitals, and home care, which support their integrated health model. The information from these systems was not consolidated on a single platform, and its access and availability generated an operative load, which obstructs all health management processes and the support of clinical decisions for physicians. Consequently, we proposed the design and implementation of a healthcare, clinical, and business data repository with advanced analytic capabilities to consume machine learning prediction models to improve the decision-making process and population health management at the organization. The digital health platform conceptual framework is shown in Figure 1.

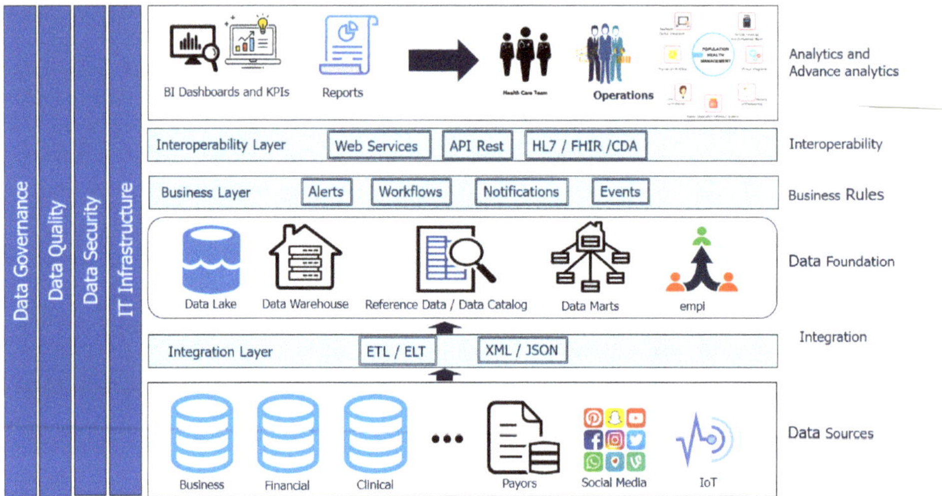

Figure 1. Conceptual Framework—Keralty Health Portal.

The implementation of the platform was an ambitious project that required integrating health information from disparate sources, building numerous technological and functional components, and the definition of IT management processes robust enough to support interoperability with other systems. The digital health information platform included patient-related data, Electronic Health Records (EHR), diagnostic reports, prescriptions, medical images, pharmacy records, research data, operational data, financial data, and human resources data.

This project was innovative and pioneered the designing and building of a comprehensive health digital platform for a healthcare organization in Colombia, with the patient being at the center of it and all of its information aggregated and summarized based on the standardized enterprise data repository. This information can be accessed quickly and intuitively when and where it is needed, hiding all technical complexity and providing longitudinal process management tools, as well as tools for decision support for professionals. The difference of this platform with other implementations was the development of a medical portal with a patient 360 view that uses data from the enterprise data repository to generate real-time early warning scores, patient surveillance, open API for hospitals integration, prediction of health risk patterns, high-risk markers, co-morbidity detection to predict critical diseases, early diagnosis of diseases, treatment comparison with medical guidelines, and measurement of efficiency of specific drugs to provide the best quality of care.

The Digital Healthcare platform architecture can ingest data from over 50 different source systems at the granular level, including claims, clinical, financial, administrative, wearables, genomics, and socioeconomic data. Few platforms today can integrate that many heterogeneous data sources successfully. The platform can consume machine learning models on-demand without the need for further development. The data logic models are on top of the raw data and can be accessed, reused, and updated through open APIs without the need for clinical and business logic changes. The platform was able to integrate successfully structured and unstructured data. It is commonly seen that this type of platforms in the market is built to either integrate structured data or unstructured but few cases successfully integrate both. Open microservices APIs were created for operations such as authorization, identity management, interoperability, and data pipeline management. These microservices enable the development of third-party applications to interoperate with the platform.

2.1. Platform Architecture

The initial approach was to build a big data processing pipeline with a Microsoft Azure lambda architecture to support real-time and batch analytics. This approach is shown in Figure 2. This architecture has different mechanisms to consume data depending on the source and timing needed to generate insights. In addition, with this approach, we can have professionals with different skills working in parallel to build the platform.

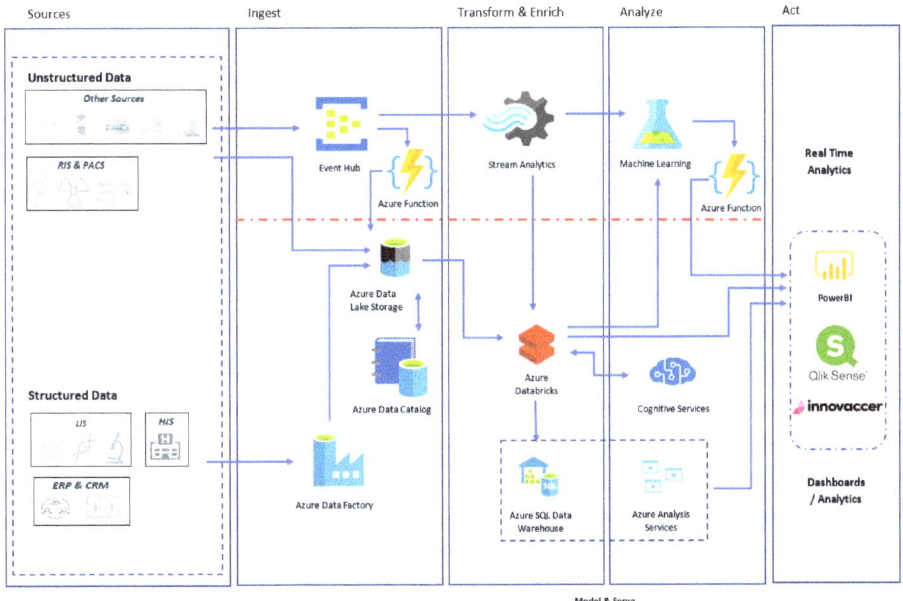

Figure 2. Azure Big Data and Machine Learning Lambda Architecture.

The architecture contains a batch layer, a real-time layer, and a serving layer. The batch layer is in charge of persistent storage and is able to scale horizontally. The real-time layer process streaming data and performs dynamic computation. The serving layer query data on the repositories and consume the prediction models.

From the infrastructure point of view, the platform offers the flexibility of being implemented in a hybrid environment, namely the cloud and the local data processing center, through the use of virtualization techniques, containers, and load balancing systems. The design of the infrastructure was prepared to provide a flexible set of resources that can be used on-demand and based on the specific workload requirements. The infrastructure deployment relied heavily on automation to provide fluid operations.

2.2. Data Repository

An enterprise-wide staging repository for the big data analytics platform was considered. The data lake allows capturing data of any volume, type, and ingestion speed in one single place for storing heterogeneous data. This staging area included capabilities such as security, scalability, reliability, and availability. The data can be passed, processed directly from the staging area, or can be ingested to an enterprise data warehouse for historical load, preparation, and serve for BI and machine learning needs. This data warehouse repository has a scale-out architecture and massively parallel processing (MPP) engine.

Data models were developed to cover clinical, social, and healthcare program domains. Each model performs validations and processing on the data received, decoupling the processing and administration of the data from the source. These data models can also be extended to store additional attributes specific to the implementation, allowing these models to subscribe to certain types of messages, using the mapping and filtering options provided by the data processing pipelines. Once these subscriptions are created, the model will be loaded with all relevant messages to those who are subscribed and stored in the data lake.

For data storage, the data are loaded into a data warehouse with a daily refresh. This healthcare data repository contains a highly normalized data model for fast and efficient querying and analysis. This repository is read-only.

2.3. Integration and Interoperability

The platform provides a mechanism to integrate data from heterogeneous sources, define workflows to ingest data from different data stores, and transform and process data to data stores to be consumed by BI applications. A cloud-based data integration service is used to create these data-driven workflows and orchestrate all automation, transformation, and data movement in the platform. The main tasks this integration service should perform are: creation and scheduling of data pipelines to ingest data from different data sources, processing and transformation of the data, and store data in data stores such as data lakes or data warehouses.

Azure Data Factory automates and orchestrates the entire data integration process from end to end in the platform. We built the ETL (extract, transform, and load) pipelines with this Azure component. The data are extracted from the source locations, transformed from its source format to the target Azure data lake's schema, and loaded into Azure data lake and the data warehouse, where they can be used for analytics and reporting. Azure Data Factory defines control flows that execute various tasks in the transform and load process.

We used the mechanism called mapping data flows, combining control flows and data flows to build the data transformations with an easy-to-use visual user interface. These data flows are then executed as activities within Azure Data Factory pipelines. Data Factory is certified by HIPAA (Health Insurance Portability and Accountability Act), which protects the data while they are in use with Azure. In the data flow, we created transformation streams where we define the source data and create the graph with the transformations, schema operations such as derived column, aggregate, surrogate keys and selects, and the output settings.

2.4. Data Security and Privacy Model

In terms of security, the platform guarantees authentication, access control, and encryption capabilities. The security mechanisms of the platform can provide protection, alert monitoring, and support the OAuth 2.0 protocol for authentication with REST interfaces. ACLs are enabled on folders, subfolders, and files. The platform also provides encryption mechanisms to protect the data.

All these capabilities are accompanied by the implementation of enterprise security policies and regulatory compliance requirements.

2.5. Stream Analytics

The platform can handle mission-critical real-time data and offer end to end streaming pipelines with continuous integration and continuous delivery (CI-CD) services. Other capabilities such as in-memory processing, data encryption, and support of international security standards including HIPAA (Health Insurance Portability and Accountability Act), HITRUST (Health Information Trust Alliance), and GDPR (General Data Protection Regulation).

2.6. Advanced Analytics

The analytic data component consists of two areas: The first area is the BI models we develop for tactical, operational, and strategic decisions. The second area comprehends several prediction models that need to be developed. Currently, there are two prediction models developed by the authors of this paper to support population health management, specifically the diagnosis of sepsis and hypertension prediction [24,25]. These insights assist providers in the detection and tracking of chronic diseases. The machine learning component is used to build, test, consume, and deploy predictive analytic models on-demand and as requested for the organization. The platform provides self-service dashboards and visualizations that use data from the repositories to drive the decision-making process. The machine learning application layer is one of the essential layers of this platform.

Once the data are integrated, aggregated, and normalized in the system, the platform offers a tool to provide knowledge management through the business intelligence interface providing data analysis, design, and training of machine learning models, as well as development and management of results-based care indicators or population health management. The platform provides a tool where clinicians, researchers, and scientists can mine the data and get valuable information.

Machine learning models can be trained and customized in preconfigured data domains, allowing the storage of the results for future use. Data researchers and scientists can develop advanced tools to obtain information and value of the data stored in the solution, taking advantage of the model design, training, and validation component. We briefly present the predictive models implemented in the platforms.

- **Machine Learning Classification for a Hypertensive Population:** This prediction model evaluates the association between gender, race, BMI (Body Mass Index), age, smoking, kidney disease, and diabetes using logistic regression. Data were collected from NHANES datasets from 2007 to 2016 to train and test the model, a dataset of 19,709 samples with (83%) non-hypertensive individuals and (17%) hypertensive individuals. The results show a sensitivity of 77%, a specificity of 68%, precision on the positive predicted value of 32% in the test sample, and a calculated AUC of 0.73 (95% CI [0.70–0.76]). The model used to estimate the probability that a person will belong to the hypertensive or non-hypertensive class is:

$$p = \frac{e^{(\beta_0+\beta_1 gender+\beta_2 age+\beta_3 race+\beta_4 bmi+\beta_5 kidney+\beta_6 smoke+\beta_7 diabetes)}}{1+e^{(\beta_0+\beta_1 gender+\beta_2 age+\beta_3 race+\beta_4 bmi+\beta_5 kidney+\beta_6 smoke+\beta_7 diabetes)}}$$

We used the logistic regression classification model in this experiment to evaluate the importance of the risk factor variables and their relationship with the prevalence of hypertension among a nationally representative sample of adults ≥20 years in the United States (n = 19,759). The distribution of the samples by hypertensive patients, gender, and race is shown in Table 1.

Table 1. Number of samples by hypertensive class, gender, and race.

Hypertension, Adults 20 and over—2007–2016			
Class	Gender	Race	n
Non Hypertensive	Female	Mexican American	1269
		Non-Hispanic Black	1674
		Non-Hispanic White	**3674**
		Other Hispanic	951
		Other Race—Including Multi-Racial	864
	Male	Mexican American	1255
		Non-Hispanic Black	1599
		Non-Hispanic White	**3714**
		Other Hispanic	774
		Other Race—Including Multi-Racial	843
Hypertensive	Female	Mexican American	205
		Non-Hispanic Black	420
		Non-Hispanic White	**662**
		Other Hispanic	149
		Other Race—Including Multi-Racial	114
	Male	Mexican American	214
		Non-Hispanic Black	478
		Non-Hispanic White	**670**
		Other Hispanic	138
		Other Race—Including Multi-Racial	132
		Total	**19,799**

We computed chi-square test between each independent variable and the dependent variable to indicate the strength of evidence that there is some association between the variables. Chi-square was selected due to the categorical form of the data used in the model, and it is considered one of the best methods to estimate the dependency between the class and the features when the feature can take a fixed number of possible values that belong to a group or nominal category.

Table 2 shows the *p*-value for each variable; the null hypothesis is reject for any $p \leq 0.05$, while the null hypothesis is not rejected when $p > 0.05$. *p*-values for the variables GENDER, BMIRANGE_1, BMIRANGE_3, and KIDNEY_2 are not statistically significant at 0.05 alpha level; the clinical importance of these variables in the model for interpretation allows us to include them. We ran the model with and without the variables, and there were no significant changes in the accuracy score, positive predicted value rate, and true positive rate.

The training dataset was derived from a random sampling of 70% (13,831) of the extracted study population and the test sampling the remaining 30% (5928) to evaluate the model on the ground-truth that was never used for training. We ran the logistic regression model on the entire dataset to verify the accuracy score of the model.

Table 2. Chi2 test and p-value for the independent variables.

Chi-Squared between Each Indicator Variable and the Baseline for the Model				
Feature	**Description**	**Dummy**	***p*-Value**	**Score**
GENDER	Male	GENDER_1	0.1416446	2.160001
	Female	GENDER_2	0.1450268	2.123795
AGERANGE	20–30	AGERANGE_1	0.0000001	560.890568
	31–40	AGERANGE_2	0.0000001	299.675698
	41–50	AGERANGE_3	0.0000001	98.221463
	51–60	AGERANGE_4	0.0000035	21.520345
	61–70	AGERANGE_5	0.0000001	342.879412
	71–80	AGERANGE_6	0.0000001	1037.137074
RACE	Mexican American	RACE_1	0.0067797	7.330429
	Other Hispanic	RACE_2	0.0275756	4.854409
	Non-Hispanic White	RACE_3	0.0455912	3.996636
	Non-Hispanic Black	RACE_4	0.0000001	91.264812
	Other Race	RACE_5	0.0000278	17.562718
BMIRANGE	Underweight = <18.5	BMIRANGE_1	0.6730361	0.178071
	Normal weight = 18.5–24.9	BMIRANGE_2	0.000033	17.234712
	Overweight = 25–29.9	BMIRANGE_3	0.9174572	0.010741
	Obesity = BMI of 30 or greater	BMIRANGE_4	0.0006362	11.666854
KIDNEY	Yes	KIDNEY_1	0.0000001	58.963059
	No	KIDNEY_2	0.1872889	1.738816
SMOKE	Yes	SMOKE_1	0.0021759	9.394891
	No	SMOKE_2	0.0053461	7.758468
DIABETES	Yes	DIABETES_1	0.0000001	217.214128
	No	DIABETES_2	0.0000001	39.351672
	Borderline	DIABETES_3	0.0000051	20.798905

The Logistic Regression model uses the logit function to express the relationship of the risk factors as:

$$logit(p) = ln(\frac{p}{1-p}) = \beta_0 + \beta_1 X_1 + \beta_2 X_2 + ... + \beta_i X_i$$

The probability of success can be expressed as:

$$p = \frac{e^{(\beta_0 + \beta_1 X_1 + \beta_2 X_2 + ... + \beta_i X_i)}}{1 + e^{(\beta_0 + \beta_1 X_1 + \beta_2 X_2 + ... + \beta_i X_i)}}$$

where p is the predicted probability of having hypertension, X_i are the risk factors or independent variables, and β_i are the coefficients that are estimated by using the method of maximum likelihood and allow us to calculate the odds that, for every unit increase in X_i, the odds of having hypertension changes by e^{β}.

- **A neural network approach to predict early neonatal sepsis:** We developed a non-invasive neural network classification model for early neonatal sepsis detection. The data used in this study are from Crecer's Hospital center in Cartagena-Colombia. A dataset of 555 neonates with (66%) of negative cases and (34%) of positive cases was used to train and test the model. The study results show a sensitivity of 80.32%, a specificity of 90.4%, precision on the positive predicted value of 83.1% in the test, sample and a calculated area under the curve of 0.925 (95% Confidence Interval [91.4–93.06]). The neural network architecture can be seen in Figure 3.

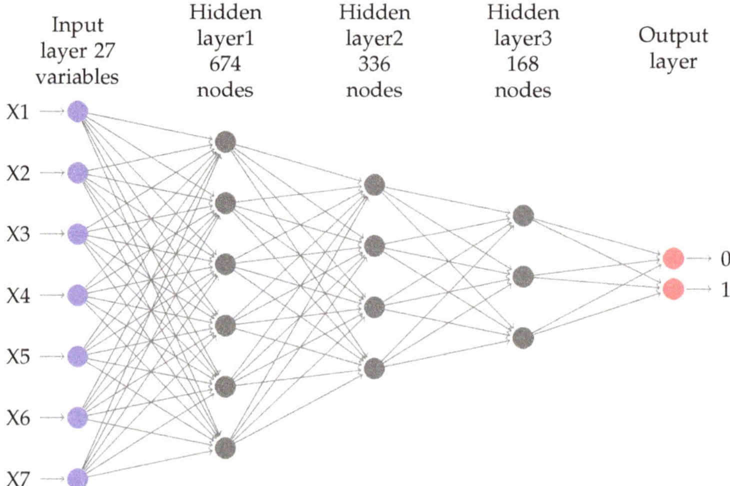

Figure 3. Multilayer Perceptron Architecture.

Table 3 shows the parameters of the architecture. Labels X1–X7 are informative only, and the input size is 27 variables.

Table 3. Model architecture parameters.

Model Architecture Parameters	
Parameter	**Value**
Input Dimension	27
Num Output classes	2
Num Hidden Layers	3
Hidden Layer1 Dimension	674
Activation Func Layer1	Relu
Hidden Layer2 Dimension	336
Activation Func Layer2	Relu
Hidden Layer3 Dimension	168
Activation Func Layer3	Relu
Minibatch size	8
Num samples to train	388
Num minibatches to train	48
Loss Function	cross entropy with softmax
Eval Error	Classification error
Learner for parameters	momentum sgd
Eval Metrics	Confusion Matrix, AUC

The model used an anonymous dataset from a private medical institution in Cartagena, Colombia, from 2016 to 2017. Demographic, laboratory data, blood pressure, and body measures data were part of the dataset. This dataset includes cases of live newborns of ages inferior to 72 h with a diagnosis of early neonatal sepsis by clinical criteria and laboratory blood cultures. Control cases were part of the dataset including all newborns healthy by clinical diagnosis and who returned healthy for a follow up at 72 h.

This retrospective study includes 186 cases and 368 controls based on a case-control relationship of 1:2 with a 95% trust factor and power of 80%. Bivariate analysis and logistic regression were performed to detect the variables associated with early sepsis, and the statistical significance was considered at the alpha level of 0.05.

This model considered nine sociodemographic, fourteen obstetric, nine neonatal, and four maternal infectious related pathology variables. Table 4 shows the quantitative sociodemographic variables, Table 5 shows the qualitative sociodemographic variables, Table 6 shows the quantitative neonatal variables, Table 7 shows the qualitative neonatal variables, Table 8 shows the quantitative obstetric variables, Table 9 shows the qualitative obstetric variables, and Table 10 shows the qualitative maternal infections of the cases and controls.

A bivariate chi-square test with correction was performed to the qualitative variables to find a statistical association between the independent variable and the possibility to develop early neonatal sepsis. For continuous variables, the Mann–Whitney U test was performed. From this statistical analysis, it is essential to show that we did not find significant statistical evidence for the variables age, start of marital status at younger than 18 years old, gender, APGAR (Appearance, Pulse, Grimace, Activity, and Respiration) value less than 7 after 1 and 5 min, the number of pregnancies, and the type of birth. Prenatal control is not associated with the case of sepsis; however, assisting to five prenatal controls are associated with the protection to avoid the appearance of early neonatal sepsis. There was no evidence with the variables IUGR (Intrauterine Growth Restriction) background and multiple pregnancies. Twenty-seven (27) variables were selected as input variables for our artificial neural network architecture.

Table 4. Quantitative sociodemographic variables in cases (186) and controls (369).

Quantitative Socio Demographic Variable	Cases				Controls				
	Mean	Median	SD	RIQ	Mean	Median	SD	RIQ	*p*-Value
Age	23.93	23.5	4.99	20–26	24.22	23	6.19	19–28	0.793
Onset of sexual activity	16.06	16	0.945	15–17	15.6	16	0.971	15–16	0.0001

Table 5. Qualitative sociodemographic variables in cases (186) and controls (369).

Qualitative Socio Demographic Variable	Categories	Cases		Controls		X2	*p*-Value
		N	%	N	%		
Teen Mother	Yes No	15 171	8.1 91.9	69 300	18.7 81.3	10.88	0.001
Health Regimen	Government Commercial	183 3	98.4 1.6	349 20	94.6 5.4	4.51	0.041
Origin	Rural Urban	42 144	22.6 77.4	5 364	1.4 98.6	71.87	0.00001
Marital Status	Married or in common law married Single, divorced or widow	128 58	68.8 31.2	101 268	27.4 72.6	87.64	0.00001
Level of education	Elementary School High School	86 100	46.2 53.8	80 289	21.7 78.3	35.57	0.00001
Start of Marital status life younger than 18 yo	Yes No	178 8	95.7 4.3	357 12	96.7 3.3	0.39	0.531
Start of Marital status life younger than 16 yo	Yes No	47 139	25.3 74.7	147 222	39.8 60.2	11.54	0.001

Table 6. Quantitative Neonatal variables in cases (186) and controls (369).

Quantitative Neonatal Variable	Cases				Controls				
	Mean	Median	SD	RIQ	Mean	Median	SD	RIQ	*p*-Value
New born weight in grams	2639.9	2768.5	546.5	2500–3020	3202.4	3224	412.1	2950–3500	0.0001
APGAR after 1 min of birth	7.73	8.0	0.611	8.0	8.09	8.0	0.598	8.0	0.0001

Table 7. Qualitative Neonatal variables in cases (186) and controls (369).

Qualitative Neonatal Variable	Categories	Cases		Controls		X2	p-Value
		N	%	N	%		
Premature	Yes No	100 86	53.8 46.2	25 344	6.8 93.2	156.4	0.0001
Gender	Male Female	109 77	58.6 41.4	202 167	54.7 45.3	0.748	1.672
Less than 1500 grams	Yes No	11 175	5.9 94.1	2 367	0.5 99.5	15.6	0.00001
Less than 2500 grams	Yes No	44 142	23.7 76.3	9 360	2.4 97.6	64.44	0.00001
APGAR less than 7 after 1 min of birth	Yes No	2 184	1.1 98.9	3 366	0.8 99.2	0.095	0.999
APGAR less than 7 after 5 min	Yes No	4 182	2.2 97.8	9 360	2.4 97.6	0.045	0.999
Respiratory distress	Yes No	89 97	47.8 52.2	27 342	7.3 92.7	122.8	0.0001

Table 8. Quantitative Obstetric variables in cases (186) and controls (369).

Quantitative Obstetric Variable	Cases				Controls				
	Mean	Median	SD	RIQ	Mean	Median	SD	RIQ	p-Value
Gestational age at the time of birth	35.6	36.0	3.47	34–39	38.4	39.0	1.62	38–39	0.0001
Number of prenatal controls	4.08	5.0	1.83	3.75–5.0	4.32	5.0	1.83	4–5.0	0.002
Number of pregnacies	1.77	1.0	1.15	1.0–2.0	1.6	1.0	1.15	1–2.0	0.076
Number of births	1.04	1.0	1.03	0–1	0.7	1.0	1.03	0–1	0.0001
Numbers of C-sections	0.65	1.0	0.68	0–1	0.76	1.0	0.68	0–1	0.029

Table 9. Qualitative Obstetric variables in cases (186) and controls (369).

Qualitative Obstetric Variable	Categories	Cases		Controls		X2	p-Value
		N	%	N	%		
Type of birth	Vaginal C-Section	98 88	52.7 47.3	162 207	43.9 56.1	3.833	0.05
IUGR Background	Yes No	5 181	2.7 97.3	13 356	3.5 96.5	0.275	0.6
Assistance for prenatal control	Yes No	165 21	88.7 11.3	318 51	86.2 13.8	0.702	0.402
Assistance for at least 4 prenatal control	Yes No	140 46	75.3 24.7	301 68	81.6 18.4	3.01	0.083
Assistance for at least 5 prenatal control	Yes No	105 81	56.5 43.5	254 115	68.8 31.2	8.301	0.004
Premature rupture of membrane with more than 18 hours	Yes No	95 91	51.1 48.9	17 352	4.6 95.4	165.7	0.00001
Chorioamnionitis	Yes No	23 163	12.4 87.6	3 366	0.8 99.2	36.96	0.00001
Premature membrane rupture with more than 6 hours	Yes No	161 25	86.6 13.4	194 175	52.6 47.4	61.96	0.0001
Multiple Pregnacies	Yes No	2 184	1.1 98.9	10 359	2.7 97.3	0.39	0.353

Table 10. Qualitative maternal infections variables in cases (186) and controls (369).

Qualitative Maternal Infections Variables	Categories	Cases		Controls		X2	p-Value
		N	%	N	%		
Maternal Fever	Yes	67	36.0	40	10.8	50.38	0.0001
	No	119	64.0	329	89.2		
Yeast Infections	Yes	31	16.7	15	4.1	25.83	0.0001
	No	155	83.3	354	95.9		
Sexualy transmitted disease history	Yes	27	14.5	7	1.9	34.24	0.0001
	No	159	85.5	362	98.1		
Urinary Tract Infections	Yes	11	5.9	9	2.4	4.29	0.0381
	No	175	94.1	360	97.6		

In terms of computational timing, It is difficult to evaluate the complexity and timing of a machine learning algorithm. However, based on the algorithmic complexity, we can measure the time performance in terms of its training time complexity using big O notation because the classification time of the models can vary depending on the stress in the computational performance and power. In terms of timing, the classification prediction with the trained models is less than 1 s. The time complexity of the logistic regression could be expressed as $O((f+1)csE)$, where f is the number of features (+1 because of bias), c is the number of possible outputs, s is the number of samples, and E is the number of epochs to run. For the neural network approach, $O(pnl_1 + nl_1nl_2 + \ldots)$, where p is the number of features and nl_i is the number of neurons at layer i in a neural network [26].

3. Actual Platform Benefits

The implementation of the platform became the digital healthcare ecosystem for the organization. The organization can populate workflow information systems with critical decision-making insights, accurate and reliable healthcare data that significantly increased the value of the healthcare outcome to patients and care providers. This platform delivers significant benefits to the organization, such as physicians having an intelligent application that can be configured to their preferences and optimized to their disciplines, patients receiving more personalized care, an improvement in healthcare workflow and patient care, and personalized care for physicians and patients.

We describe in the following subsections several use cases that effectively present the change and digital transformation of the organization with the implementation of the platform.

3.1. Reduce Total Cost of Care for Care Coordination

With a robust data analytic component, the organization was able to prioritize opportunities for improvement and to improve the way care is coordinated and delivered throughout its network of hospitals and medical facilities. The results include a considerable increase in financial results in just six months.

The organization uses the platform to generate timely, meaningful, and actionable data to drive change and improve the quality of care for patients. The organization uses the data for risk-stratification of the network's population, prioritization of the care coordination activities, and prevention activity's interventions. Risk stratification was completed for all patients, enabling care managers to identify individuals at various risk levels for unnecessary services and high-cost utilization, improving patient outcomes and experience. The analytical component also reduces unnecessary visits, facilitates access to specialty care and community-based services, and achieves healthcare outcomes. Other benefits include 3% increase in the detection of high-risk patients with primary care, 20% increase in the number of patients with ongoing care managed, and 10% percent reduction in emergency department utilization per member among care managed patients.

3.2. Self-Service Analytic

As described in this paper, the healthcare platform combines and standardizes data across different source systems to provide actionable insights in a single platform. The platform integrates data from different sources, such as claims data, cost data, financial data, clinical data, and other patient data. With self-service analytics, the organization increases the number of users accessing the analytic component, improving data visibility and providing actionable insights to improve patient outcomes.

3.3. Reduce Deaths from Sepsis

The organization improved sepsis mortality rates and improving care outcomes by using the advanced analytic component of the platform. Sepsis impacts almost 1.7 million adults in the U.S. and is responsible for nearly 270,000 annual deaths. One-third of all hospital deaths are patients with sepsis [27]. The machine learning prediction model used in the platform was developed by one of the authors of this paper, as described before. It is still too early to mention the results of the utilization of this feature. However, the goal of the organization is to reduce its sepsis mortality rate, the costs of the creation of its sepsis care transformation team, and the implementation of an evidence-based sepsis care practice.

3.4. Discussion and Limitations

The digital health platform helps Keralty organization with closing the gaps between multiple datasets, improving clinical benefits, improving patient's lives, supporting better decision-making to manage larger populations, and improving overall health outcomes. However, the need for algorithms with high accuracy in medical diagnosis is still a challenge that needs to be improved precisely and efficiently [28]. The increasing complexity of building end-to-end platforms to integrate disparate systems and to apply machine learning techniques in specific areas such as computer vision, natural language processing, reinforcement learning, and other generalized methods present many challenges when forming the interdisciplinary team needed and the set of technological components used for the implementation.

Some challenges should be considered in the design and implementation of machine learning projects for healthcare. One of the most critical challenges requires algorithms that can answer causal questions. These questions are beyond classical machine learning algorithms because they require a formal model of interventions [29]. To address this type of question from the analytical component of the platform, we need to learn from data differently and to gain knowledge in causal models to understand how machine learning algorithms need to be trained. Another challenge is to create reliable outcomes from heterogeneous data sources with the participation of SME (Subject Matter Experts) who understand the disease; the machine learning predictive accuracy and correct clinical interpretation depend on the criteria and context of the disease. Providers and machine learning engineers should work together on model interpretability and applicability. Machine learning implementation is not an easy task; the selection of predictive features and optimization of hyperparameters is another challenge that needs to be mastered to implement models that provide useful insights [30]. The success and meaningful use of these algorithms, and their integration into the platform depends on the accuracy of the models and their interpretability.

4. Results of Advanced Analytics

After training and testing the logistic regression model for predicting hypertension, we generated some evaluation metrics to evaluate the classifier. Table 11 shows the confusion matrix with the classification results, include the true positive value (730), true negative value (3407), false negative (216), and false positive value (1575). The classification report in Table 12 shows the calculated precision and sensitivity.

Table 11. Confusion matrix.

		Predicted	
		Non-Hypertensive	**Hypertensive**
True	**Non-Hypertensive**	3407	1575
	Hypertensive	216	730

Table 12. Classification report.

Classification Report				
	Precision	**Recall**	**f1-Score**	**Support**
Non-Hypertensive	0.94	0.68	0.79	4982
Hypertensive	0.32	0.77	0.45	946
avg/total	0.84	0.7	0.74	5928

The test sampling of 5928 contains 4982 (84%) non-hypertensive and 946 (16%) hypertensive patients. The model shows a sensitivity of 730/946 = 77% and a specificity of 3407/4982 = 68%. The precision of the model was 730/2305 = 32% and the negative predicted value 3407/3623 = 94%. The false negative rate of the model was 216/946 = 22%. The model was better at identifying individuals who will not develop hypertension than those who will develop hypertension.

For the neural network approach to predict early neonatal sepsis, Table 13 shows the confusion matrix with the classification results of actual class label vs. the predicted ones, including the true positive value (49), true negative value (95), false negative (12), and false positive value (10).

Table 13. Confusion matrix.

		Predicted	
		Non-Sepsis	**Sepsis**
True	**Non-Sepsis**	95	10
	Sepsis	12	49

The classification report in Table 14 shows the precision and sensitivity. The sensitivity of the model is moderately acceptable due to the imbalanced testing dataset, and there is still a high number of false negatives.

Table 14. Classification report.

Classification Report			
True Positive	**False Negative**	**Precision**	**Accuracy**
49	12	0.83	0.867
False Positive	**True Negative**	**Recall**	**f1-score**
10	95	0.803	0.817
Positive Label: 1		**Negative Label: 0**	

A sensitivity of 80.3% and a specificity of 90.4% show that the model might be useful for detecting positive cases, and the true negative rate shows that the model is also efficient at identifying negative cases. The high precision value of 83.1% and the AUC of 0.925 confirm the adequacy of the model as a preliminary screening tool. The percentage of positive cases shows that the model works better than random guessing and the conditional probability of negative test results is considerably low.

The accuracy of 86.74% shows that the model correctly identifies negative cases and positive cases based on the characteristics of the dataset and the small number of cases examined.

5. Comparison with Other Platforms

A review of several healthcare platforms shows that the architecture presented in this paper covers all the categories from integration, interoperability, security care, and advanced analytics. Generally, other implementations only focused on one specific area, as shown in Table 15 and taken from the International Conference on Computational Intelligence and Data Science (ICCIDS 2018) and a healthcare frameworks review proposed in the Journal of King Saud University [31].

Table 15. Comparison of healthcare big data platforms.

Author and Year	Patient Centric	Predictive Analysis	Real Time Monitoring	Improve Treatment	Interoperability	Workflow and Rules	Pop Health	Patient 360
Our Health Platform	Yes	Yes	Yes	Yes	Yes	Yes	Yes	Yes
Raghupathi et al. (2014) [32]	Yes	No	Yes	Yes	Partial	No	Partial	No
Patel et al. (2016) [33]	Yes	Yes	Yes	Yes	Partial	Partial	Yes	No
Chawla et al. (2013) [34]	Yes	Yes	No	Yes	Partial	Partial	Yes	No
Abinaya et al. (2015) [35]	Partial	Yes	Partial	Yes	Yes	Partial	No	No
Balladini et al. (2015) [36]	Yes	No	Yes	Yes	Partial	Yes	No	No
Belle et al. (2015) [37]	Partial	No	Yes	Yes	Partial	Yes	No	No
Mezghani et al. (2015)	Partial	No	Yes	Yes	Yes	Partial	No	No
Chen et al. (2017) [38]	Yes	Partial	Yes	Yes	Yes	Partial	Yes	No

We designed and implemented a healthcare platform using big data technologies with actionable insights to augment human decision-making at the organization impacting the population's health, public health, and to capture social determinants of health. This platform comprehends all the features we use in the comparison. Raghupathi et al. reported a conceptual architecture to present big data analytic outlines in healthcare with no predictive analytic capabilities and no patient 360 view. Patel et al. designed a big data architecture platform to improve data aggregation in the healthcare industry and to provide a reduction in healthcare cost, predicting analytic, preventive care, and drug discovery capabilities but without patient 360 view capabilities. Chawla et al. presented a patient-centric healthcare framework—Collaborative Assessment and Recommendation Engine (CARE)—to improve patient-centric treatment and diagnosis without real-time monitoring and 360 view capabilities. Abinaya et al. implemented a fascinating e-Health service application for diagnosing heart diseases. Balladini et al. designed a real-time architecture of big data for Francisco Lopez Lima Hospital in Argentina to process physiological data. This platform did not include predictive analytic and patient 360 view. Belle et al. implemented a genomic data processing platform that provides image analytic and signal processing of psychological data. Mezghani et al. designed a big data platform for integrating heterogeneous wearable data in healthcare for real-time monitoring and diagnosis. Lastly, Chen et al. presented a real-time big data platform to improve communication and collaboration between patients and providers, increasing the quality of care that clinical teams can provide.

6. Conclusions and Future Work

This paper provides details of an optimized and secure healthcare platform that revolutionizes the healthcare industry in Colombia by providing better information to patients and care teams. The use of this technology reduces the costs associated with healthcare.

The proposed digital health platform allows us to address population health challenges, to understand better patient's health, and to find hidden patterns that traditional data analytics fail to find. The organization can use unified patient-generated data, financial data, and socioeconomic data to detect patterns and to discover a group of patients who share similar health behavior. The analysis of clinical and non-clinical data allows predicting patient's health with better accuracy. The platform also allows better health discoveries and actions based on treatment history for individuals and groups of patients.

Keralty organization recognized that better care coordination was required for patients receiving care. The organization wanted to improve quality outcomes, provider engagement and recruitment, and its own economic health. To meet these objectives, the organization focuses on clinician engagement and organizational alignment, ensuring widespread access to meaningful, actionable data, and the use of the healthcare analytics platform to inform decisions and drive improvement. Keralty believes the use of machine learning will be one of the most important, life-saving technologies ever introduced to the organization. We believe the opportunities are virtually limitless for the platform to improve and accelerate clinical, workflow, and financial outcomes.

More future work needs to be done on the platform to continue improving all the benefits for the entire organization. Tools for performing knowledge discovery process will be added to the ecosystem. The organization is planning to start the implementation of prescriptive analytics models to assist the organization in making smarter decisions in population health management. The architecture team will look at the possibility of implementing Map/Reduce-based computations for processing data with high scalability and to execute low latency and high concurrency analytical queries on top of Hadoop clusters.

Author Contributions: Conceptualization, F.L.-M., V.G.-D. and E.R.N.-V.; Methodology, F.L.-M., V.G.-D. and E.R.N.-V.; Software, F.L.-M.; Validation, F.L.-M., V.G.-D., Z.B. and E.R.N.-V.; Formal Analysis, F.L.-M.; Investigation, F.L.-M., V.G.-D., Z.B. and E.R.N.-V.; Resources, F.L.-M.; Data Curation, F.L.-M. and Z.B.; Writing—Original Draft Preparation, F.L.-M., V.G.-D. and E.R.N.-V.; Writing—Review and Editing, F.L.-M., V.G.-D., Z.B. and E.R.N.-V.; Visualization, F.L.-M., Z.B. and E.R.N.-V.; Supervision, F.L.-M. and V.G.-D.; Project Administration, V.G.-D. and E.R.N.-V.; Funding Acquisition, F.L.-M. All authors have read and agreed to the published version of the manuscript.

Funding: This research received no external funding.

Acknowledgments: This document presents an independent study supported by the company Sanitas USA. The points of view expressed are those of the authors and not necessarily those of Sanitas USA. We thank Ivan Murcia VP of Healthcare Services at Sanitas USA and Santiago Thovar, CIO at Keralty who provided insight and expertise that greatly assisted the study.

Conflicts of Interest: The authors declare no conflict of interest. The founders had no role in the design of the study; in the collection, analyses, or interpretation of data; in the writing of the manuscript, or in the decision to publish the results.

Abbreviations

The following abbreviations are used in this manuscript:

ACL	Access Control List
BI	Business Intelligence
CRM	Customer Relationship Management
EHR	Electronic Health Record
ERP	Enterprise Resource Planning
GDPR	General Data Protection Regulation
HIS	Hospital Information System
HIPAA	Health Insurance Portability and Accountability Act
HITRUST	Health Information Trust Alliance
LIS	Lab Information System
MPP	Massive Parallel Computing
RIS	Radiology Information System
REST	Representational State Transfer

References

1. Glassman, A.; Giuffrida, A.; Escobar, M.L.; Giedion, U. Chapter 1 Colombia: After a Decade of Health System Reform. In *From Few to Many*; Inter-American Development Bank: Washington, DC, USA, 2009; Volume 1, pp. 1–13.
2. Ruíz, F.; Gaviria, A.; Norman, J. Plan Decenal de Salud Pública. *Bogotá* **2020**, in press.
3. Legido, H.; Lopez, P.A.; Balabanova, D.; Perel, P.; Lopez-Jaramillo, P.; Nieuwlaat, R.; Schwalm, J.D.; McCready, T.; Yusuf, S.; McKee, M. Patients' knowledge, attitudes, behaviour and health care experiences on the prevention, detection, management and control of hypertension in Colombia: A qualitative study. *PLoS ONE* **2015**, *10*, e122112. [CrossRef]
4. Lopez, F.E.; Bonfante, M.C.; Arteta, I.G.; Baldiris, R.E. IoT and big data in public health: A case study in Colombia. In *Protocols and Applications for the Industrial Internet of Things*; IGI Global: Hershey, PA, USA, 2018; pp. 309–321, ISBN 978-1-5225-3806-6.
5. Dennis, R.J.; Caraballo, L.; García, E.; Rojas, M.X.; Rondon, M.A.; Pérez, A.; Aristizabal, G.; Peñaranda, A.; Barragan, A.M.; Ahumada, V. Prevalence of asthma and other allergic conditions in Colombia 2009–2010: A cross-sectional study. *BMC Pulm. Med.* **2012**, *12*, 12. [CrossRef]
6. About Keralty. Available online: https://www.keralty.com/en/about-keralty (accessed on 27 January 2020).
7. León, G.R. Digitalización de Historia Clínica. Available online: https://contrataciondelestado.es/wps/wcm/connect/3236c434-7ce1-484f-bb50-b8942bdc7d66/DOC20190314132936Estandar_digitalizacion_SACYL-+9.pdf?MOD=AJPERES (accessed on 27 January 2020).
8. Zhang, Y.; Qiu, M.; Tsai, C.W.; Hassan, M.M.; Alamri, A. Health-CPS: Healthcare cyber-physical system assisted by cloud and big data. *IEEE Syst. J.* **2017**, *11*, 88–95. [CrossRef]
9. Mezghani, E.; Exposito, E.; Drira, K.; Da Silveira, M.; Pruski, C. A Semantic Big Data Platform for Integrating Heterogeneous Wearable Data in Healthcare. *J. Med. Syst.* **2015**, *39*. [CrossRef]
10. Kaur, P.; Sharma, M.; Mittal, M. Big Data and Machine Learning Based Secure Healthcare Framework. *Proc. Procedia Comput. Sci.* **2018**, *132*, 1049–1059. [CrossRef]
11. Thota, C.; Sundarasekar, R.; Manogaran, G.; Varatharajan, R.; Priyan, M.K. Centralized Fog Computing security platform for IoT and cloud in healthcare system. In *Fog Computing: Breakthroughs in Research and Practice*; IGI Global: Hershey, PA, USA, 2018; pp. 365–378, ISBN 978-1-5225-5650-3.
12. Edet, R.; Afolabi, B. Prospects and Challenges of Population Health with Online and other Big Data in Africa. *Adv. J. Soc. Sci.* **2019**, *6*, 57–63. [CrossRef]
13. MedAware—Using AI to Eliminate Prescription Errors—Digital Innovation and Transformation. Available online: https://digital.hbs.edu/platform-digit/submission/medaware-using-ai-to-eliminate-prescription-errors/ (accessed on 8 March 2020).
14. Ling, Z.J.; Tran, Q.T.; Fan, J.; Koh, G.C.H.; Nguyen, T.; Tan, C.S.; Yip, J.W.L.; Zhang, M. GEMINI: An integrative healthcare analytics system. *Proc. VLDB Endow.* **2014**, *7*, 1766–1771. [CrossRef]
15. Manogaran, G.; Thota, C.; Lopez, D.; Vijayakumar, V.; Abbas, K.M.; Sundarsekar, R. *Big Data Knowledge System in Healthcare*; Springer: Cham, Switzerland, 2017; pp. 133–157. [CrossRef]
16. Bates, D.W.; Saria, S.; Ohno-Machado, L.; Shah, A.; Escobar, G. Big data in health care: Using analytics to identify and manage high-risk and high-cost patients. *Health Affairs* **2014**, *33*, 1123–1131. [CrossRef]
17. Farooqi, M.M.; Shah, M.A.; Wahid, A.; Akhunzada, A.; Khan, F.; ul Amin, N.; Ali, I. Big Data in Healthcare: A Survey. *Appl. Intell. Technol. Healthc.* **2019**, 143–152. [CrossRef]
18. Hulsen, T.; Jamuar, S.S.; Moody, A.R.; Karnes, J.H.; Varga, O.; Hedensted, S.; Spreafico, R.; Hafler, D.A.; McKinney, E.F. From big data to precision medicine. *Front. Media* **2019**, *6*, 34. [CrossRef]
19. Hatzigeorgiou, M.N.; Joshi, M.S. Population Health Systems: The Intersection of Care Delivery and Health Delivery. *Popul. Health Manag.* **2019**, *22*, 467–469. [CrossRef]
20. Koti, M.S.; Alamma, B.H. Predictive analytics techniques using big data for healthcare databases. In *Proceedings of the Smart Innovation, Systems and Technologies*; Springer Science and Business Media: Singapore, 2019; Volume 105, pp. 679–686. [CrossRef]
21. Dash, S.; Shakyawar, S.K.; Sharma, M.; Kaushik, S. Big data in healthcare: Management, Analysis and Future prospects. *J. Big Data* **2019**, *6*, 54. [CrossRef]
22. Puaschunder, J.M. Big Data, Algorithms and Health Data. *SSRN Electron. J.* **2019**. [CrossRef]

23. Moreira, M.W.; Rodrigues, J.J.; Korotaev, V.; Al-Muhtadi, J.; Kumar, N. A Comprehensive Review on Smart Decision Support Systems for Health Care. *Inst. Electr. Electron. Eng.* **2019**, *13*, 3536–3545. [CrossRef]
24. López-Martínez, F.; Núñez-Valdez, E.R.; Lorduy Gomez, J.; García-Díaz, V. A neural network approach to predict early neonatal sepsis. *Comput. Electr. Eng.* **2019**, *76*, 379–388. [CrossRef]
25. López-Martínez, F.; Schwarcz.MD, A.; Núñez-Valdez, E.R.; García-Díaz, V. Machine Learning Classification Analysis for a Hypertensive Population as a Function of Several Risk Factors. *Expert Syst. Appl.* **2018**, *110*, 206–215. [CrossRef]
26. Singh, A. Foundations of Machine Learning. *SSRN Electron. J.* **2019**, *486*. [CrossRef]
27. Rhee, C.; Dantes, R.; Epstein, L.; Murphy, D.J.; Seymour, C.W.; Iwashyna, T.J.; Kadri, S.S.; Angus, D.C.; Danner, R.L.; Fiore, A.E.; et al. Incidence and trends of sepsis in US hospitals using clinical vs claims data, 2009–2014. *JAMA J. Am. Med. Assoc.* **2017**, *318*, 1241–1249. [CrossRef]
28. Mahindrakar, P.; Hanumanthappa, M. Data Mining In Healthcare: A Survey of Techniques and Algorithms with Its Limitations and Challenges. *Int. J. Eng. Res. Appl.* **2013**, *3*, 937–941.
29. Ghassemi, M.; Naumann, T.; Schulam, P.; Beam, A.L.; Chen, I.Y.; Ranganath, R. A Review of Challenges and Opportunities in Machine Learning for Health 2018. *arXiv* **2018**, arXiv:1806.00388.
30. Waring, J.; Lindvall, C.; Umeton, R. Automated machine learning: Review of the state-of-the-art and opportunities for healthcare. *Artif. Intell. Med.* **2020**, *104*, 101822. [CrossRef]
31. Palanisamy, V.; Thirunavukarasu, R. Implications of big data analytics in developing healthcare frameworks—A review. *J. King Saud Univ. Comput. Inf. Sci.* **2019**, *31*, 415–425. [CrossRef]
32. Raghupathi, W.; Raghupathi, V. Big data analytics in healthcare: Promise and potential. *Health Inf. Sci. Syst.* **2014**, *2*. [CrossRef] [PubMed]
33. Patel, S.; Patel, A. A Big Data Revolution in Health Care Sector: Opportunities, Challenges and Technological Advancements. *Int. J. Inf. Sci. Tech.* **2016**, *6*, 155–162. [CrossRef]
34. Chawla, N.V.; Davis, D.A. Bringing big data to personalized healthcare: A patient-centered framework. *J. Gen. Intern. Med.* **2013**, *28*. [CrossRef] [PubMed]
35. Abinaya, K. Data Mining with Big Data e-Health Service Using Map Reduce. *IJARCCE* **2015**, *4*, 123–127. [CrossRef]
36. Balladini, J.; Rozas, C.; Frati, F.; Vicente, N.; Orlandi, C. Big Data Analytics in Intensive Care Units: Challenges and applicability in an Argentinian Hospital. *J. Comput. Sci. Technol.* **2015**, *15*, 61–67.
37. Belle, A.; Thiagarajan, R.; Soroushmehr, S.M.R.; Navidi, F.; Beard, D.A.; Najarian, K. Big data analytics in healthcare. *BioMed Res. Int.* **2015**, *2015*. [CrossRef]
38. Chen, D.; Chen, Y.; Brownlow, B.N.; Kanjamala, P.P.; Arredondo, C.A.G.; Radspinner, B.L.; Raveling, M.A. Real-time or near real-time persisting daily healthcare data into HDFS and elasticsearch index inside a big data platform. *IEEE Trans. Ind. Inform.* **2017**, *13*, 595–606. [CrossRef]

Article

Diagnosis in Tennis Serving Technique

Eugenio Roanes-Lozano [1,*], Eduardo A. Casella [2], Fernando Sánchez [3] and Antonio Hernando [4]

1 Instituto de Matemática Interdisciplinar (IMI) & Departamento de Didáctica de las Ciencias Experimentales, Sociales y Matemáticas, Facultad de Educación, Universidad Complutense de Madrid, c/Rector Royo Villanova s/n, 28040 Madrid, Spain

2 Automóvil Club San Nicolás, Chiclana 109 bis, 2900 San Nicolás, Prov. de Buenos Aires, Argentina; topocasella@gmail.com

3 Departamento de Matemáticas, Facultad de Ciencias, Universidad de Extremadura, Avda. de Elvas s/n, 06006 Badajoz, Spain; fsanchez@unex.es

4 Departamento de Sistemas Informáticos, ETSI de Sistemas Informáticos, Universidad Politécnica de Madrid, Carretera de Valencia km 7, 28031 Madrid, Spain; antonio.hernando@upm.es

* Correspondence: eroanes@mat.ucm.es; Tel.: +34-91-3946248

Received: 18 March 2020; Accepted: 22 April 2020; Published: 25 April 2020

Abstract: Tennis is a sport with a very complex technique. Amateur tennis players have trainers and/or coaches, but are not usually accompanied by them to championships. Curiously, in this sport, the result of many matches can be changed by a small hint like 'hit the ball a little higher when serving'. However, the biomechanical of a tennis stroke is only clear to an expert. We, therefore, developed a prototype of a rule-based expert system (RBES) aimed at an amateur competition player that is not accompanied by his/her coach to a championship and is not serving as usual (the RBES is so far restricted to serving). The player has to answer a set of questions about how he/she is serving that day and his/her usual serving technique and the RBES obtains a diagnosis using logic inference about the possible reasons (according of the logic rules that have been previously given to the RBES). A certain knowledge of the tennis terminology and technique is required from the player, but that is something known at this level. The underlying logic is Boolean and the inference engine is algebraic (it uses Groebner bases).

Keywords: rule-based expert systems; tennis hitting technique; computer algebra systems; Groebner bases; Boolean logic

---

1. Introduction

The first and third authors of this paper are veteran tennis players, regularly taking part in both veterans and open tennis championship (they were the +55 doubles champions of Extremadura region in 2018). They have a patent (Fernando María Sánchez Fernández, Eugenio Roanes Lozano: Dispositivo de entrenamiento para determinar la posición óptima en el tenis. Country: Spain. Request number: 201730051. Priority date: 3 de febrero de 2017. Patent holder: Universidad de Extremadura y Universidad Complutense de Madrid.) regarding the recovery position in tennis [1]. The first author has been the +45 champion of Extremadura region several times and is a tennis instructor ranked not so long ago among the top 500 players of Spain).

The second author was their tennis coach. He is an "ITF Level 2 Tennis Coach" with a long experience in teaching tennis and coaching tennis players both in Argentina and Spain.

Finally, the fourth author is an experienced expert systems developer.

Tennis is a sport with a sophisticated technique [2–4]. The big distance between the hand and the racket's sweet spot makes it difficult to play well. The high precision required is normally achieved through the learning of the technique and a hard training. Something similar happens with golf.

It is well known that there are many aspects in tennis teaching that have to be addressed (tactic, technique, physical training, etc.) and a coach is clearly needed for improving the player's level. For instance, the biomechanical of a stroke is only clear to the expert's eyes (although there are some personalized computer studies of this aspect for elite athletes like high jumpers).

Most amateur tennis players are not accompanied by a coach when playing championships, except when playing team competitions. In this kind of championship, only the captain of the team can talk to players during the matches. An experience of many tennis players is that a simple hint like "hit the ball at a higher point when you serve" can change the result of a match.

We would like to address here just this aspect of the whole process of tennis coaching: finding out what is going wrong in a certain match with respect to the player's usual technique. It applies inference to the answers of the player to a set of questions regarding both the problem and the player's usual technique. It requires a certain knowledge of the technique from the player (let us underline that it is focused to amateur competition players, not to beginners). Therefore, this is neither a system devoted to learn how to play tennis, nor a substitute for a coach. It only tries to correct the mistakes of the stroke going wrong along one match and return the situation to the usual one.

The article is illustrated with the part of the system corresponding to the serve.

While the algebraic model for rule-based expert systems (RBES) is well known (see Section 4), the application to tennis is, as far as we know, entirely novel.

The expert system we have developed tries to model a well-known international standard describing the tennis serving technique. An important issue in an expert system is the knowledge base and the formalism used to represent it. Since propositional Boolean logic is enough to represent this standard, we have adopted this logic to embody the knowledge of our expert system. Artificial Intelligence has proposed many elaborated representation formalisms to model uncertainty or fuzzy knowledge in expert systems, as well as machine learning techniques to acquire this knowledge base. However, we have not required all these techniques for designing our system because of the same purpose of our system: output the recommendations for tennis serving technique, according to a well-known international standard. For our purpose we have only needed to model this standard by means of propositional Boolean logic, resulting in a system with a high performance.

2. Some Introductory Notes about Rule-Based Expert Systems Based on Boolean Propositional Logic

In this section, we will describe some outlines of RBES based on Boolean propositional logic for representing knowledge. It may be skipped by an acquainted reader, but it is included in order to make the article self-contained, as it could have readers from different environments.

Boolean propositional logic uses a finite set of atomic propositions $X_1, \ldots, X_m$ for defining formulae (Definition 1) through the connectives $\neg$, $\vee$, $\wedge$, and $\rightarrow$.

Definition 1 (Formula). *A formula A is recursively defined as:*

- *X_i where X_i is an atomic proposition*
- *$\neg B$ where B is a formula*
- *$B \wedge C$ where B, C are formulae*
- *$B \vee C$ where B, C are formulae*
- *$B \rightarrow C$ where B, C are formulae.*

We define $\mathcal{C}$ as the set of formulae.

Definition 2 (Rule). *By its importance, we define rules as the formulae with the form $(A_1 \wedge A_2 \wedge \ldots \wedge A_k) \rightarrow (A_{k+1} \vee \ldots \vee A_n)$ where each formula $A_1, \ldots, A_n$ is either an atomic proposition (X) or the negation of an atomic proposition ($\neg X$).*

Remark 1. *We have chosen the previous definition of rule because we believe it is the most intuitive one. A rule of the form:*

$B \to C \vee D$

is equivalent to the rule:

$B \wedge \neg C \to D$

but it is less intuitive as what is a cause and what is an effect is not on different sides of the implication symbol. Similarly, the implications could be substituted by disjunctions, what has the advantage of using fewer symbols, but would make it less readable.

Remark 2. *Sometimes some rules with the same antecedent and different consequents are grouped into a single formula for the sake of brevity and clarity. For example, the information in the three rules:*

$B \to C$
$B \to D$
$B \to E$

can be summarized in the formula $B \to C \wedge D \wedge E$. Although it is not formally correct (it doesn't comply with Definition 2), we shall hereinafter admit these abbreviations as rules too (in order to save space and to clarify the underlying ideas related to tennis hitting technique). The algebraic approach used has no problem to deal with this sort of formulae, and the timings obtained for globally computing which output are obtained are negligible (as will be shown in Section 6).

Definition 3 (IC). *An* integrity constraint (IC) *is a piece of knowledge added by the experts and expressing that some potential facts cannot hold at the same time.*

Definition 4 (elements of a RBES). *By means of the concept of formula, we can define the elements of which a RBES is composed:*

Input: *The input of a RBES is concerned with the information related to the environment of the RBES. This information is described by means of a finite number of different atomic propositions.*
Output: *The output of a RBES is concerned with the information inferred by the RBES. It is also described by means of a finite number of atomic propositions.*
Knowledge-Base: *The knowledge-base of the RBES is concerned with the information contained in the system, which is used along with the input of the RBES to infer the output of the system. In a RBES based on propositional Boolean logic, this knowledge-base will be mostly represented by a finite set of rules which may require to define new atomic propositions. Consequently, the knowledge-base of the RBES is described by a finite set of formulae $\{R_1, ..., R_{n'}\} \subset C$.*

The concept of tautological consequence of propositional logic is used for determining the knowledge which the RBES can infer. As usual, we make use of $\{A_1, ..., A_n\} \models B$ to denote that the formula B is a tautological consequence of the formulae $A_1, ..., A_n$ (we say that $\{A_1, ..., A_n\} \models B$ if and only if, whenever the formulae $A_1,...,A_n$ hold, then B also holds). In this way, a RBES with knowledge base $\{R_1, ..., R_{n'}\}$ and having as input the set of formulae $\{F_1, ..., F_n\}$ infers the formula $B \in C$ if and only if the following holds:

$$\{F_1, ..., F_n, R_1, ..., R_{n'}\} \models B$$

Another important issue on propositional Boolean logic is related to the concept of consistency (in an informal way, $\{A_1, ..., A_n\}$ is inconsistent if both a formula and its negation can be simultaneously inferred). Obviously, the formulae embodying the knowledge-base of a RBES must be consistent.

3. Organizing the Knowledge of the Tennis Hitting Technique RBES

The propositional variables of the RBES have been grouped in four homogeneous blocks:

- Facts – Block I (Generalities): kind of serve the player is executing and player's level (variables x_i, y_i, z_i are used in this block).
- Facts – Block II (Details): details of the execution of the serve, kinetic and coordination.
- Guards: intermediate conclusions—not visible for the user (variables g_i are used for the guards).
- Output: conclusions about the way the player is serving (variables c_i are used for the output).

The structure of the RBES is summarized in Figure 1.

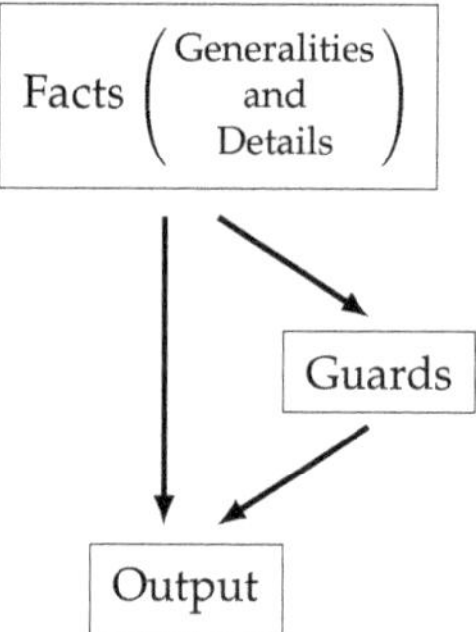

Figure 1. Some rules directly deduce output from the facts. Other rules lead to intermediate results (guards). And another rules obtain output from the guards. Therefore, the figure shows the hierarchy among input variables (facts), intermediate conclusions (guards) and final conclusions (output).

3.1. *Facts—Block I (Generalities)*

In this first block consists just of three questions: the system is informed about the kind of serve the player is executing and two technical details about his/her abilities (see Table 1). We shall briefly explain these facts for those unaware of tennis technique.

Table 1. Facts—Block I.

Description	Propositional Variable
Kind of serve (choose exactly one):	
You are executing a slice serve	x_1
You are executing a flat serve	x_2
You are executing a topspin serve	x_3
You are an expert player and you can serve in any of the three ways (x_1, x_2, x_3) tossing the ball the same place	z_0
You usually jump when executing this kind of serve	y_1

Regarding the spin:

- in a flat serve the ball is not rotating,
- in the slice serve it rotates around a vertical axis (the trajectory is curved and bounces to one side),
- in the topspin serve the ball is rotating around an horizontal axis (the upper part of the ball moves forward and the lower part moves backwards): the aerodynamic effect is that the ball bounces closer and higher.

For the sake of simplicity we consider x_1, x_2, x_3 mutually exclusive. This is not completely accurate, as an advanced player could serve with a mixture of topspin and slice spins.

Each kind of serve normally requires the ball to be tossed at different places (see l_1, l_2, l_3 below) (Let us underline that the encoding chosen use redundant variables, that could be avoided if using 3-valued variables instead. Nevertheless, according to our experience, Boolean computations are much faster, even if more variables are needed.). But this way the opponent is aware of the kind of spin of the serve. It is really difficult although possible to always toss the ball at the same place (z_0) (Observe that there is no particular reason for choosing the subscripts of the variables (0, 1, 2, 3, 4, 5, 6, 7) the way they have been chosen.). Roger Federer is a good example of this ability.

If the serve is correctly executed, the player tends to move upwards (y_1) and forward (inside the court).

3.2. Facts—Block II (Details)

Tennis is not a precise science in the sense that there isn't a unique valid truth for everyone. There are different "schools" and ways to teach and even to execute the different strokes.

They also change with time: before Björn Borg nearly no player played two-handed backhand and used topspin. Nowadays the vast majority of professional players play two-handed backhand and use topspin in most strokes from the back of the court. Meanwhile, two-handed forehand wasn't a success and was only used by a few players like Hans Gildemeister.

There are very successful unorthodox styles, like Jim Courier's baseball-style forehand or Rafael Nadal's forehand finishing. A system like the one presented here would be useless in such a case.

There are also great tennis players with an untraditional serve, like John P. McEnroe, and players with a pure traditional service, like Roger Federer.

Therefore we have just considered what could be considered the standard serving technique [2–4], and have transcribed the causality underlying what is explained in the tennis books into logical rules to the best of our knowledge.

After many years of development of this sport, we consider that there is a consensus in the clear cause-effect relation in tennis hitting technique. For instance, a clear simplified example is the following: if you serve and the ball is too far away in front of you and too low when you hit it, the racket will be facing the floor when it hits the ball, so the ball will bounce just in front of you, never on the other side of the net. Obviously most situations are more complex, but the consensus in the standard technique and the above mentioned causality is what makes plausible (from our point of view) to use a Boolean logic approach, as we compare what the player is doing with the standard theoretical technique (the latter admitted as a truth). Something due to the authors is how they have translated that consensual knowledge into rules.

As said in Section 3.2 above, we have considered that the different spins are mutually exclusive as simplification.

Tables 2 and 3 (splitted only for the sake of space) collects the most typical items of the execution of the serve to be taken into account (in our personal opinion).

Let us underline the difference between t_1 and a_1: the toss can be correct or even too high, but the player can let the ball fall too low anyway.

It is clear that we have the following incompatibilities:

- d_1 and d_2 are incompatible,
- t_1, t_2 and t_3 are mutually incompatible,
- l_1, l_2 and l_3 are mutually incompatible,
- a_1 and a_2 are incompatible.

Table 2. Facts—Block IIa.

Description	Propositional Variable
Feet initial position / distance (choose at most one):	
Distance between feet too big	d_1
Distance between feet too small	d_2
Arms initial situation:	
Arms initially not relaxed	v_1
Shoulders' initial position (angle w.r.t. baseline) close to right angle (or your usual position)	w_1
Tossing the ball—A (choose exactly one):	
Too low toss	t_1
Normal height toss	t_2
Too high toss	t_3
Tossing the ball—B (choose exactly one):	
A bit towards the arm holding the racket	l_1
On the middle	l_2
A bit towards the arm not holding the racket	l_3
Hitting the ball (choose at most one):	
You are letting the ball drop too low	a_1
You are hitting the ball too high (upper part of the racket)	a_2
Looking at the ball:	
You are keeping your eyes on the ball	m_1
Grip:	
You are holding the grip firmly	e_1
This is your usual grip for this kind of serve	e_2

Table 3. Facts—Block IIb.

Description	Propositional Variable
Kinetic:	
You flex slightly your knees before hitting the ball	k_1
Your knees are straight when hitting the ball	k_2
You balance slightly backward and forward during the execution of the serve and you move or tend to move forward	k_3
You feel your abdominals do work during the execution of the serve	k_4
Your racket begins the swing more or less behind your head at its usual initial position	k_5
The swing of your racket finishes too early	k_6
You are hitting the ball with your feet on the air	k_7
Coordination:	
You "feel coordinated" when serving today	o_1

3.3. Guards

Some rules will clearly have an immediate consequence in the serve, but others just provoke intermediate consequences that we have denoted "guards" (Table 4). We have decided to introduce these technical intermediate steps in the rule base in order to organize it more clearly, as they somehow give technical reasons for the knowledge expressed in the rules involved. They can be very useful for an expert in tennis and logic revising the RBES.

Table 4. Guards.

Description	Propositional Variable
Lack of balance (equilibrium of the body)	g_1
Physical difficulty or impossibility to execute the stroke that way	g_2
Acceleration w.r.t. normal rhythm or some parts of the execution of the serve not on time	g_3
Incompatibility with the player's own technique	g_4

Initially, we have decided that guards were not visible to the players, as they are technical reasons for reaching an effect in the serve. Nevertheless they could be an explanation for a player that is curious for obtaining a logic explanation of what happens, as they are somehow an intermediate level of diagnosis. If it is considered interesting for the system to include them as visible output, it would be enough to compute some more normal forms (see Sections 6.1 and 6.2 below).

3.4. Output

Let us detail the main different output of performing a certain serve (Table 5).

Table 5. Output.

Description	Propositional Variable
Imprecision	c_1
Stroke too weak (lack of power)	c_2
Too low and/or too short	c_3
Too long	c_4
Little spin	c_5

No incompatibility should be included among the output. Different technical errors could lead to opposite outputs, what is not contradictory.

3.5. Rules and Incompatibilities (Integrity Constraints)

We shall not detail all the rules but we shall begin by giving an example of two simple concatenated rules.

Rule 5 is:

$$R5 : t1 \to g1$$

and it means:

R5: Too low toss $\to$ Lack of balance (equilibrium of the body)

(the consequent not an output, but a guard—an intermediate technical conclusion, not visible for the end user). This rule expresses that, if the player throws the ball in to the air too low, he/she will hit the ball in a flexed position, what will provoke a lack of balance.

Rule 23 is:

$$R23 : g1 \to c1 \wedge c2$$

and it means:

R23: Lack of balance (equilibrium of the body) $\to$ Imprecision $\wedge$ Stroke too weak (lack of power) (the latter are conclusions—output). Therefore, if we have $t1$, we conclude $c1$ and $c3$ This is a very simple example, just to get the flavour of the process. We do not detail all rules in this way for the sake of space.

The rules regarding the facts considered are:

$R1 : d1 \rightarrow c2$
$R2 : d2 \rightarrow g1$
$R3 : v1 \rightarrow c2$
$R4 : \neg w1 \rightarrow g1$
$R5 : t1 \rightarrow g1$
$R6 : t3 \rightarrow c1$
$R7 : l1 \wedge \neg z0 \wedge \neg x1 \rightarrow g2 \wedge c5$
$R8 : l2 \wedge \neg z0 \wedge \neg x2 \rightarrow g2 \wedge c5$
$R9 : l3 \wedge \neg z0 \wedge \neg x3 \rightarrow g2 \wedge c5$
$R10 : a1 \rightarrow c3 \; xor \; c4$
$R11 : a2 \rightarrow g3$
$R12 : \neg m1 \rightarrow c1$
$R13 : \neg e1 \rightarrow c1 \wedge c2$
$R14 : \neg e2 \rightarrow g4$
$R15 : \neg k1 \rightarrow c1 \wedge c2$
$R16 : \neg k2 \rightarrow c1 \wedge c2$
$R17 : \neg k3 \rightarrow c1 \wedge c2$
$R18 : \neg k4 \rightarrow c1 \wedge c2$
$R19 : \neg k5 \rightarrow c1 \wedge c2$
$R20 : k6 \rightarrow c2$
$R21 : \neg k7 \wedge y1 \rightarrow c2 \wedge c5$
$R22 : \neg e1 \rightarrow c1 \wedge c2$

and the rules corresponding to the guards are:

$R23 : g1 \rightarrow c1 \wedge c2$
$R24 : g2 \rightarrow c1 \wedge c2$
$R25 : g3 \rightarrow c1$
$R26 : g4 \rightarrow c1$

Let us underline that we have grouped some rules for the sake of brevity. For instance

$R7 : l1 \wedge \neg z0 \wedge \neg x1 \rightarrow g2 \wedge c5$

should be formally written

$R7a : l1 \wedge \neg z0 \wedge \neg x1 \rightarrow g2$
$R7b : l1 \wedge \neg z0 \wedge \neg x1 \rightarrow c5$

(this way it would be in accordance with Definition 4).

Similarly, we have included xor (exclusive or) in $R10$:

$R10 : a1 \rightarrow c3 \; xor \; c4$

The knowledge behind this rule could have also be written:

$R10n : a1 \rightarrow c3 \; \vee \; c4$
$IC0 : \neg(c3 \wedge c4)$

(note that in this case $R10n \wedge IC0 \rightarrow R10$ but $R10 \not\rightarrow R10n \wedge IC0$).

Next, we enumerate integrity constraints related to choose one option over three ones:

$IC1 : \neg(x1 \wedge x2)$
$IC2 : \neg(x2 \wedge x3)$
$IC3 : \neg(x1 \wedge x3)$
$IC4 : x1 \vee x2 \vee x3$
$IC5 : \neg(t1 \wedge t2)$
$IC6 : \neg(t2 \wedge t3)$
$IC7 : \neg(t1 \wedge t3)$
$IC8 : t1 \vee t2 \vee t3$
$IC9 : \neg(l1 \wedge l2)$
$IC10 : \neg(l2 \wedge l3)$
$IC11 : \neg(l1 \wedge l3)$
$IC12 : l1 \vee l2 \vee l3$

and the integrity constraints regarding choosing at most one option out of two are:

$IC13 : \neg(a1 \wedge a2)$
$IC14 : \neg(d1 \wedge d2)$

4. An Algebraic Model for RBES

In this section, we will see how we can implement an expert system based on Boolean propositional logic in a computer algebra system.

Let us consider an expert system based on Boolean propositional logic with propositions $X_1 \ldots X_m$. Each formula in $\mathcal{C}$ can be represented as a Boolean polynomial (see Definition 5). This representation makes use of the ideal I in $\mathbb{Z}_2[x_1, ..., x_m]$:

$$I = \langle x_1^2 - x_1, ..., x_m^2 - x_m \rangle$$

Notation 1 (NF). *Let $NF(pol, I)$ denote the 'normal form' of polynomial pol modulo ideal I.*

Definition 5 (function φ). *We define recursively the function $\varphi : \mathcal{C} \longrightarrow \mathbb{Z}_2[x_1, ..., x_m]$ as follows:*

- *If $A \equiv X_i$, then $\varphi(A) = x_i$*
- *If $A \equiv \neg B$ where $B \in \mathcal{C}$, then $\varphi(A) = NF(1 + \varphi(B), I)$*
- *If $A \equiv B \wedge C$ where $B, C \in \mathcal{C}$, then $\varphi(A) = NF(\varphi(B) \cdot \varphi(C), I)$*
- *If $A \equiv B \vee C$ where $B, C \in \mathcal{C}$, then $\varphi(A) = NF(\varphi(B) + \varphi(C) + \varphi(B) \cdot \varphi(C), I)$*
- *If $A \equiv B \rightarrow C$ where $B, C \in \mathcal{C}$, then $\varphi(A) = NF(1 + \varphi(B) + \varphi(B) \cdot \varphi(C), I)$.*

As may be seen in the previous definition, for any formula A, $\varphi(A)$ is a polynomial in $\mathbb{Z}_2[x_1, ..., x_m]$ whose variables are never to a power greater than 1 (continuously computing the reduction modulo ideal I is equivalent to work in the residue class ring $\mathbb{Z}_2[x_1, ..., x_m]/I$). Besides, the total degree of the polynomial may be, at most, equal to the number of variables.

Once described how Boolean formulae are represented by polynomials, we show, by Theorem 1, how the problem of determining if a formula is tautological consequence of others can be translated into an algebraic problem [5]. Previous works not providing a residue class ring as a model for the logic and not considering the extension to RBES are [6,7] (Boolean case) and [8,9] (many-valued case). A detailed survey can be found in [10]. There are recent developments in this line of research such as [11,12].

Theorem 1. *Let $A_1, ..., A_n, B \in \mathcal{C}$. The following holds:*

(i) $\{A_1, ..., A_n\} \models B \Leftrightarrow \varphi(\neg B) \in \langle \varphi(\neg A_1), ..., \varphi(\neg A_n) \rangle + I$
(ii) $\{A_1, ..., A_n\}$ *is consistent* $\Leftrightarrow 1 \notin \langle \varphi(\neg A_1), ..., \varphi(\neg A_n) \rangle + I$

Theorem 1 is important since the algebraic problems involved in this theorem may be solved making use of Groebner bases [13–15]. On the ground of this, many expert systems have been developed in different fields [16–21].

Indeed, the question of determining whether a RBES with knowledge-base $\{R_1, ..., R_{n'}\}$ infers the formula B when its input is the set of facts $\{F_1, ..., F_n\}$ may be solved by checking whether the following holds:

$$NF(\varphi(\neg B), I + \langle \varphi(\neg F_1), ..., \varphi(\neg F_n), \varphi(\neg R_1), ..., \varphi(\neg R_n) \rangle) = 0$$

We can simplify the previous expression through the definition of the following ideals:

$$J = \langle \varphi(\neg F_1), ..., \varphi(\neg F_n) \rangle$$

$$K = \langle \varphi(\neg R_1), ..., \varphi(\neg R_{n'}) \rangle$$

In this way, the expression above results into:

$$NF(\varphi(\neg B), I + J + K) = 0$$

Similarly, the consistency of such RBES for the set of facts $\{F_1, ..., F_n\}$ can be decided using Groebner bases, as it is equivalent to:

$$GB(I + \langle \varphi(\neg F_1), ..., \varphi(\neg F_n), \varphi(\neg R_1), ..., \varphi(\neg R_n) \rangle) \neq \langle 1 \rangle$$

what can be written:

$$GB(I + J + K) \neq \langle 1 \rangle$$

5. *Maple* Implementation

The implementation used has been written in the computer algebra system *Maple* (*Maple* is a trademark of Waterloo Maple Inc., Waterloo, ON, Canada) and was introduced in 2008 [22] . The complete code is included in order to show how brief the code of the algebraic approach is (just a few lines). Firstly the algebraic packages for computing Groebner bases and normal forms and for declaring the polynomial ring where calculations will take place, are loaded:

```
restart;
with(Groebner):
with(Ore_algebra):
```

Afterwards the list of polynomial variables, the polynomial ring, and the ordering (pure lexicographic with the order for variables given by list `SV`), are declared:

```
SV:=x1,x2,x3,z0,y1,d1,d2,v1,w1,t1,t2,t3,l1,l2,l3,a1,a2,m1,e1,e2,
    k1,k2,k3,k4,k5,k6,k7,o1,g1,g2,g3,g4,c1,c2,c3,c4,c5:
A:=poly_algebra(SV,characteristic=2):
Orde:=MonomialOrder(A,'plex'(SV)):
```

Now the ideal (`iI`), generated by the square of the polynomial variables minus themselves, is defined using the auxiliary function `fu`:

```
fu:=var->var^2-var:
iI:=map(fu,[SV]):
```

Finally, the logical connectives are defined as functions (the binary ones have an "&" in order to be infix operators) and are algebraic expressions reduced modulo the ideal `iI` in ring `A` using the ordering `OrdeiI`:

```
 NEG    :=(m::algebraic) -> NormalForm(1+'m',iI,Orde):
'&AND' :=(m::algebraic,n::algebraic) ->
             NormalForm(expand(m*n),iI,Orde):
'&OR'  :=(m::algebraic,n::algebraic) ->
             NormalForm(expand(m+n+m*n),iI,Orde):
'&IMP' :=(m::algebraic,n::algebraic) ->
             NormalForm(expand(1+m+m*n),iI,Orde):
'&XOR' :=(m::algebraic,n::algebraic) ->
             (m &OR n) &AND NEG(m &AND n):
```

The translation of the rules and integrity constraints of the RBES developed in this article in the algebraic approach detailed above is:

```
R1:= d1 &IMP c2:
R2:= d2 &IMP g1:
R3:= v1 &IMP c2:
R4:= NEG(w1) &IMP g1:
R5:= t1 &IMP g1:
R6:= t3 &IMP c1:
R7:= (l1 &AND NEG(z0) &AND NEG(x1)) &IMP (g2 &AND c5):
R8:= (l2 &AND NEG(z0) &AND NEG(x2)) &IMP (g2 &AND c5):
R9:= (l3 &AND NEG(z0) &AND NEG(x3)) &IMP (g2 &AND c5):
R10:= a1 &IMP (c3 &XOR c4):
R11:= a2 &IMP g3:
R12:= NEG(m1) &IMP c1:
R13:= NEG(e1) &IMP (c1 &AND c2):
R14:= NEG(e2) &IMP g4:
R15:= NEG(k1) &IMP (c1 &AND c2):
R16:= NEG(k2) &IMP (c1 &AND c2):
R17:= NEG(k3) &IMP (c1 &AND c2):
R18:= NEG(k4) &IMP (c1 &AND c2):
R19:= NEG(k5) &IMP (c1 &AND c2):
R20:= k6 &IMP c2:
R21:= (NEG(k7) &AND y1) &IMP (c2 &AND c5):
R22:= NEG(e1) &IMP (c1 &AND c2):
R23:= g1 &IMP (c1 &AND c2):
R24:= g2 &IMP (c1 &AND c2):
R25:= g3 &IMP c1:
R26:= g4 &IMP c1:
IC1:= NEG(x1 &AND x2):
IC2:= NEG(x2 &AND x3):
IC3:= NEG(x1 &AND x3):
IC4:= x1 &OR x2 &OR x3:
IC5:= NEG(t1 &AND t2):
IC6:= NEG(t2 &AND t3):
IC7:= NEG(t1 &AND t3):
IC8:= t1 &OR t2 &OR t3:
IC9:= NEG(l1 &AND l2):
IC10:= NEG(l2 &AND l3):
IC11:= NEG(l1 &AND l3):
IC12:= l1 &OR l2 &OR l3:
IC13:= NEG(a1 &AND a2):
```

```
IC14:= NEG(d1 &AND d2):
```

And the (polynomial) ideal of rules is, in this case:

```
J:=[ NEG(R1),  NEG(R2),  NEG(R3),  NEG(R4), NEG(R5), NEG(R6),
     NEG(R7),  NEG(R8),  NEG(R9),  NEG(R10),NEG(R11),NEG(R12),
     NEG(R13), NEG(R14), NEG(R15), NEG(R16),NEG(R17),NEG(R18),
     NEG(R19), NEG(R20), NEG(R21), NEG(R22),NEG(R23),NEG(R24),
     NEG(R25), NEG(R26), NEG(IC1), NEG(IC2),NEG(IC3),NEG(IC4),
     NEG(IC5), NEG(IC6), NEG(IC7), NEG(IC8),NEG(IC9),NEG(IC10),
     NEG(IC11),NEG(IC12),NEG(IC13),NEG(IC14) ]:
```

Let us observe that the generators of the ideals are given as lists of polynomials. The Groebner bases of the ideals are automatically computed by *Maple* command `Basis` just by introducing the list of generators and the chosen order, as shown in Section 6.

6. Examples

6.1. A Correct Serve

Let us introduce $x3$, $\neg z0$, $y1$, $\neg d1$, $\neg d2$, $\neg v1$, $w1$, $t2$, $l3$, $\neg a1$, $\neg a2$, $m1$, $e1$, $e2$, $k1$, $k2$, $k3$, $k4$, $\neg k6$, $k7$, $o1$ as facts. The ideal `K` of facts will be:

```
K:=[ NEG(x3),NEG(NEG(z0)),NEG(y1),NEG(NEG(d1)),NEG(NEG(d2)),
     NEG(NEG(v1)),NEG(w1),NEG(t2),NEG(l3),NEG(NEG(a1)),
     NEG(NEG(a2)),NEG(m1),NEG(e1),NEG(e2),NEG(k1),NEG(k2),
     NEG(k3),NEG(k4),NEG(NEG(k6)),NEG(k7),NEG(o1) ]:
```

and the Groebner basis of the ideal generated by the ideal $I + J + K$ (note that I is denoted `iI` in the *Maple* implementation) will be:

```
time0:=time():
B:=Basis([op(iI),op(J),op(K)], Orde);
                         [c5^2 c5,...,x2,x1]
time()-time0;
                                     0.296
```

(the complete output is omitted for he sake of space). This ideal is not $\langle 1 \rangle$, so there is no inconsistency (for these facts).

Let us check if any of $c1$, $c2$, $c3$, $c4$, $c5$ follow from these facts:

```
time0:=time():
NormalForm(NEG(c1),B,Orde);
                                     1 + c1
NormalForm(NEG(c2),B,Orde);
                                     1 + c2
NormalForm(NEG(c3),B,Orde);
                                     c3 + 1
NormalForm(NEG(c4),B,Orde);
                                     1 + c4
NormalForm(NEG(c5),B,Orde);
                                     c5 + 1
time()-time0;
                                     0.078
```

As 0 was never obtained, none of the c_i is obtained by forward firing. It can be seen above how the answers were computed in less than one-tenth of a second (on a standard PC), after the previous computation of the Groebner basis of the ideal $I + J + K$ (what took less than three-tenths of a second).

6.2. An Incorrect Serve

Let us exchange now $l3$ by $l1$. Then the ball is not correctly tossed according to the serve chosen, as the player has affirmed that he/she cannot serve in any of the three ways tossing the ball the same place ($\neg z_0$)

The ideal K of facts will now be:

```
K:=[ NEG(x3),NEG(NEG(z0)),NEG(y1),NEG(NEG(d1)),NEG(NEG(d2)),
     NEG(NEG(v1)),NEG(w1),NEG(t2),NEG(l1),NEG(NEG(a1)),
     NEG(NEG(a2)),NEG(m1),NEG(e1),NEG(e2),NEG(k1),NEG(k2),
     NEG(k3),NEG(k4),NEG(NEG(k6)),NEG(k7),NEG(o1) ]:
```

and the Groebner basis of the ideal generated by $I + J + K$ will be:

```
time0:=time():
B:=Basis([op(iI),op(J),op(K)], Orde):
                          [c5+1,...,x2,x1]
time()-time0;
                                0.218
```

(the complete output is again omitted for he sake of space). This ideal is not $\langle 1 \rangle$, so there is no inconsistency (for these facts) either.

Let us check if any of $c1$, $c2$, $c3$, $c4$, $c5$ follow from these facts:

```
tiempo:=time():
NormalForm(NEG(c1),B,Orde);
                                  0
NormalForm(NEG(c2),B,Orde);
                                  0
NormalForm(NEG(c3),B,Orde);
                                c3 + 1
NormalForm(NEG(c4),B,Orde);
                                c4 + 1
NormalForm(NEG(c5),B,Orde);
                                  0
time()-time0;
                                0.078
```

Therefore c_1, c_2 and c_5 are obtained by forward firing, that is, we could have:

- imprecision
- stroke too weak (lack of power)
- little spin.

The answers were computed in similar times.

6.3. A Remark about Timings

Although it is well-known that the complexity of Groebner bases computation is double exponential in the worst case in the general case, there are very important simplifications in the algebraic model for RBES knowledge extraction and verification:

- the base field of the polynomial ring is $\mathbb{Z}_2$ (it has characteristic 2, instead of 0),
- the degree in each variable is at most 1 for all polynomials,
- the total degree of all polynomials is less or equal to the number of variables

(it is a polynomial Boolean ring, or, if the structure of the RBES is directly translated, a polynomial Boolean algebra [10]).

These features allow to treat problems of a size that would not be treatable, for instance, in real geometry. For example, the RBES of [18] consists of 313 rules (of the type specified in Remark 1) that are simplified to 182 complex rules (in the way mentioned in Remark 2). The problem treated here, with 26 rules and 14 integrity constraints, is tiny in comparison.

The specialized software PolyBoRi [23] is extremely fast due to dealing only with Boolean rings, that is, with this special kind of rings.

7. Conclusions

The main achievement is that there is no comparable RBES devoted to tennis hitting technique (as far as we know).

This is a first step in the exploration of this kind of RBES. Many more things could be done.

As future work, a comfortable GUI should be developed, either stand alone, like in [18], or accessible through Internet, like [19]. If *Maple* was the chosen tool to implement the RBES, *Maplets* could be used, although it would be even faster to use a specialized software such as MiniSat [24] or PolyBoRi [23]. Nevertheless, performance is not the key issue here, as we are just exploring the field and the system is (at least so far) simple.

Another important extension is to include backward reasoning in order to automatically determine which single propositional changes could change a certain output. This can be achieved with the same algebraic approach used, as shown in [25], but it development is not trivial and is left for a future work.

Also the possibility to use modal many-valued logics is not yet discarded, as the player can have doubts and perhaps not all variables should be considered Boolean.

Obviously the system could be detailed with more facts (like wind) and more complex possibilities (like mixing slice and topspin effects). Moreover, it could be extended to the other main strokes of tennis.

With the implementation included above, the player can study the effect of changing some of the given facts.

Author Contributions: Theoretical mathematical and computational details of the approach: E.R.-L., F.S., A.H.; underlying tennis concepts and ideas and subsequent logical rules: E.R.-L., E.A.C., F.S. All authors have read and agree to the published version of the manuscript.

Funding: This research was partially funded by the research project PGC2018-096509-B-100 (Government of Spain).

Acknowledgments: We would like to thank the anonymous reviewers for their mot valuable comments and suggestions, that have greatly improved the article.

Conflicts of Interest: The authors declare no conflict of interest.

Abbreviations

The following abbreviations are used in this manuscript:

MDPI	Multidisciplinary Digital Publishing Institute
DOAJ	Directory of open access journals
RBES	Rule Base Expert System

References

1. Roanes-Lozano, E.; Sánchez, F. An Educational Application of Dynamic Geometry: Revisiting the "Recovery Position" in Tennis. *Int. J. Tech. Math. Educ.* **2017**, *24*, 171–178.
2. Crespo, M.; Miley, D. *Advanced Coaches Manual*; ITF Ltd.: London, UK, 1998.
3. Forti, U. *Curso de Tenis*; Ed. de Vecchi: Barcelona, Spain, 1991.

4. Forti, U. *Curso Avanzado de Tenis*; Ed. de Vecchi: Barcelona, Spain, 1992.
5. Roanes-Lozano, E.; Laita, L. M.; Roanes-Macías, E. A Polynomial Model for Multivalued Logics with a Touch of Algebraic Geometry and Computer Algebra. *Math. Comp. Simul.* **1998**, *45*, 83–99.
6. Hsiang, J. Refutational Theorem Proving using Term-Rewriting Systems. *Art. Intell.* **1985**, *25*, 255–300.
7. Kapur, D.; Narendran, P. An Equational Approach to Theorem Proving in First-Order Predicate Calculus. In Proceedings of the 9th International Joint Conference on Artificial Intelligence (IJCAI-85), Los Angeles, CA, USA, 18–23 August 1985; Volume 2, pp. 1146–1153.
8. Alonso, J.A.; Briales, E. Lógicas Polivalentes y Bases de Gröbner. In *Actas del V Congreso de Lenguajes Naturales y Lenguajes Formales*; Martin, C., Ed.; Universidad de Sevilla: Seville, Spain, 1995; pp. 307–315.
9. Chazarain, J.; Riscos, A.; Alonso, J. A.; Briales, E. Multivalued Logic and Gröbner Bases with Applications to Modal Logic. *J. Symb. Comput.* **1991**, *11*, 181–194.
10. Roanes-Lozano, E.; Laita, L.M.; Hernando, A.; Roanes-Macías, E. An Algebraic Approach to Rule Based Expert Systems. *Rev. R. Acad. Cien. Ser. A. Mat.* **2010**, *104*, 19–40.
11. Alonso-Jiménez, J.A.; Aranda-Corral, G.A.; Joaquín Borrego-Díaz, J.; Fernández-Lebrón, M.M.; Hidalgo-Doblado, M.J. A logic-algebraic tool for reasoning with Knowledge-Based Systems, *J. Log. Algebr. Meth. Program.* **2018**, *101*, 88–109.
12. Aranda-Corral, G.A.; Borrego-Díaz, J.; Fernández-Lebrón, M.M. Conservative Retractions of Propositional Logic Theories by Means of Boolean Derivatives: Theoretical Foundations. In *Intelligent Computer Mathematics. CICM 2009.*; Carette, J., Dixon, L., Coen, C.S., Watt, S.M., Eds.; Lect. Notes Comput. Sci.; Springer: Berlin/Heidelberg, Germany, 2009; Volume 5625, pp. 45–58.
13. Buchberger, B. Bruno Buchberger's PhD thesis 1965: An algorithm for finding the basis elementals of the residue class ring of a zero dimensional polynomial ideal. *J. Symb. Comput.* **2006**, *41*, 475–511.
14. Cox, D.A.; Little, J.; O'Shea, D. *Ideals, Varieties, and Algorithms: An Introduction to Computational Algebraic Geometry and Commutative Algebra*, 3rd. ed.; Undergraduate Texts in Mathematics; Springer: Berlin/Heidelberg, Germany, 2007.
15. Winkler, F. *Polynomial Algorithms in Computer Algebra, Texts and Monographs in Symbolic Computation*; Springer: Wien, Austria; New York,NY, USA, 1996.
16. Laita, L.M.; Roanes-Lozano, E.; Maojo, V.; de Ledesma, L.; Laita, L. An Expert System for Managing Medical Appropriateness Criteria Based on Computer Algebra Techniques. *Comp. Math. Appl.* **2000**, *51*, 473–481.
17. Lourdes Jimenez, M.L.; Santamaría, J.M.; Barchino, R.; Laura Laita, L.; Laita, L.M; González, L.A.; Asenjo, A. Knowledge representation for diagnosis of care problems through an expert system: Model of the auto-care deficit situations. *Exp. Syst. Appl.* **2008**, *34*, 2847–2857,
18. Pérez-Carretero, C.; Laita, L. M.; Roanes-Lozano, E.; Lázaro, L.; González-Cajal, J.; Laita, L. A Logic and Computer Algebra-Based Expert System for Diagnosis of Anorexia. *Math. Comput. Simul.* **2002**, *58*, 183–202.
19. Roanes Lozano, E.; Galán-García, J.L.; Aguilera-Venegas, G. A Portable Knowledge Based System for Car Breakdown Evaluation. *Appl. Math. Comput.* **2015**, *267*, 758–770.
20. Roanes Lozano, E.; Galán-García, J.L.; Aguilera-Venegas, G. A prototype of a RBES for personalized menus generation. *Appl. Math. Comput.* **2017**, *315*, 615–624.
21. Rodríguez-Solano, C.; Laita, L.M.; Roanes-Lozano, E.; López-Corral, L.; Laita, L. A Computational System for Diagnosis of Depressive Situations. *Exp. Syst. Appl.* **2006**, *31*, 47–55.
22. Roanes-Lozano, E.; Laita, L.M.; Roanes-Macías, E. A Groebner Bases Based Many-valued Modal Logic Implementation in Maple. In *AISC/Calculemus/MKM 2008*; Autexier, S., Campbell J., Rubio J., Sorge V., Suzuki M., Wiedijk F., Eds.; Lecture Notes in Artificial Intelligence, 5144; Springer: Berlin/Heidelberg, Germany, 2008; pp. 170–183.
23. Available online: http://polybori.sourceforge.net/ (accessed on 10 April 2020).
24. Available online: http://minisat.se/ (accessed on 10 April 2020).
25. Roanes-Lozano, E.; Hernando, A.; Laita, L.M.; Roanes-Macías, E. A Groebner bases-based approach to backward reasoning in rule based expert systems. *Ann. Math. Art. Int.* **2009**, *56*, 297–311.

Article

Decision Support System for Fitting and Mapping Nonlinear Functions with Application to Insect Pest Management in the Biological Control Context

Ritter A. Guimapi [1,2,*], Samira A. Mohamed [1], Lisa Biber-Freudenberger [3], Waweru Mwangi [2], Sunday Ekesi [1], Christian Borgemeister [3] and Henri E. Z. Tonnang [1]

1 International Centre of Insect Physiology and Ecology (ICIPE), Nairobi P.O. Box 30772-00100, Kenya; sfaris@icipe.org (S.A.M.); sekesi@icipe.org (S.E.); htonnang@gmail.com (H.E.Z.T.)

2 Department of Computing, School of Computing & Information Technology, Jomo Kenyatta University of Agriculture and Technology (JKUAT), Nairobi P.O. Box 62000-00200, Kenya; waweru_mwangi@icsit.jkuat.ac.ke

3 Center for Development Research (ZEF), Department of Ecology and Natural Resources Management, University of Bonn, Walter-Flex-Str. 3, 53113 Bonn, Germany; lfreuden@uni-bonn.de (L.B.-F.); cb@uni-bonn.de (C.B.)

* Correspondence: ritteryvan@gmail.com

Received: 4 March 2020; Accepted: 16 April 2020; Published: 24 April 2020

Abstract: The process of moving from experimental data to modeling and characterizing the dynamics and interactions in natural processes is a challenging task. This paper proposes an interactive platform for fitting data derived from experiments to mathematical expressions and carrying out spatial visualization. The platform is designed using a component-based software architectural approach, implemented in R and the Java programming languages. It uses experimental data as input for model fitting, then applies the obtained model at the landscape level via a spatial temperature grid data to yield regional and continental maps. Different modules and functionalities of the tool are presented with a case study, in which the tool is used to establish a temperature-dependent virulence model and map the potential zone of efficacy of a fungal-based biopesticide. The decision support system (DSS) was developed in generic form, and it can be used by anyone interested in fitting mathematical equations to experimental data collected following the described protocol and, depending on the type of investigation, it offers the possibility of projecting the model at the landscape level.

Keywords: Nonlinear regression; interactive platform; component-based approach; software architecture; Eclipse-RCP (Rich Client Platform); spatial prediction

1. Introduction

The ability to make reliable predictions from data through mathematical and computational concepts is fundamental in scientific research. Analytical methods expressed with mathematical modeling and simulations are often used to make predictions [1]. However, some group of scientists like biologists and entomologists are not always equipped with the necessary knowledge of calculus allowing them to perform certain types of analysis. This justifies why the various algorithms developed for fitting data to mathematical equations are not used by many. Among the techniques for fitting data to mathematical equations, nonlinear regression represents one of the most used approaches [2]. It is a very helpful process in engineering, agricultural, and natural science, and it is used to capture and understand the underling relationships among variables (dependent and independent) of interest described by mathematical expressions.

The selection of mathematical expressions for the fitting process should not be done randomly, and it has to obey a certain logic. The best way is usually guided by the theory and knowledge from the field of interest that provided the data. This insight guides the selection of the mathematical expressions used for describing the theoretical knowledge and establishing the boundaries conditions. If the mathematical expressions of the functions are too complex, it will be difficult and maybe impossible to proceed with parameter estimation. The approach should be grounded with an analytical and realistic understanding of the phenomena, which sometimes may obey biological and physical science principles. The commonly used methods for nonlinear parameter estimation include the ordinary least squares and the maximum likelihood methods [3].

The input data for model fitting may be obtained from biological and environmental processes, and it may be useful to link the resulting model to climatic variables such as temperature for spatial visualization. This approach is necessary as it allows expanding the prediction from mathematical models to be used at the landscape level as a guide in large-scale decision-making. Indeed, maps are very useful to visualize the variability of the real relationships that exist between objects and components abstracted by models. A simple model will result in a less precise map projection, while a complex model, which incorporates several variables and parameters, will allow depicting the specifics in the map. The type of model to adopt depends on its usefulness with regard to the purpose of the study and the scale to visualize [4]. However, producing a meaningful output for the end-user represents another major challenge. Simulation outputs sometimes need a special type of representation to ease their interpretation by managers who generally do not have the data analytics skills required to benefit from the proposed solution.

Spatial data are available in raster or vector format. The intrinsic nature of raster format makes them more suitable for mathematical modeling and spatial analysis. The creation of raster outputs during the mapping process is usually achieved through spatial interpolation of model projections on point locations [5,6]. There are many available interpolation approaches, which can be classified into deterministic or statistical approaches. Deterministic approaches include techniques such as proximity interpolation and inverse distance weighted (IDW), while statistical approaches include techniques such as trend surfaces and ordinary kriging [7]. Regardless of the chosen approach, the output is a map in raster format that can support agro-ecological zoning (AEZ) landscapes organized in units with similar characteristics related to the level of suitability for species or organism breeding [8]. The strategy includes the mapping of AEZ based on the inventories of climate similarities and an evaluation of the land suitability of each zone for managerial policies. In the context of biological control strategy, the concept can be adopted to identify the suitable locations for the field application of fungal-based biopesticide or a release of parasitoids of predators to reduce the population density of insect pests.

To help the end-user exploit both the fitting and the mapping features and make a complete decision, we opted to embed these features in an interactive computer-based platform that will assist uploading experimental data and support the analysis required at different stages. Such platforms are generally viewed as decision-making software (DM software) and, in most cases, are based on multi-criteria decision-making (MCDM) [9,10]. One of the purposes of DM software is indeed to provide the user with technical details, allowing them to mainly focus on the decisional aspect supported by software outputs [10]. Although there are many MCDM tools available, none of them are applicable to the type of data fitting presented here [9,11]. We take into consideration many criteria which are embedded in sequential steps for a rigorous and robust analysis.

A decision support system (DSS) is made of four major components: (1) the data management component, (2) the model management component, (3) the knowledge management component, and (4) the user interface management component [12–14]. The steps in the analysis processes are computerized to reduce the time and human effort required for making decisions [15,16]. In general, there are two types of DSS, namely, model-based and data-based systems. The model-based system is a standalone platform, which is not connected to any other information system. It requires the incorporation of mathematical expressions and applies optimization algorithms for fitting data to the expressions; it usually operates through a user interface, which ensures easy operation [17]. In contrast, the data-based system is a platform to explore and analyze large datasets, from various sources stored in databases and warehouses, before employing data mining and analytics to process the information, thereby yielding outputs [18]. The primary goal of such a platform is to support the user in decision-making for real and concrete problems.

This paper presents the design and implementation of an interactive computer-based tool for fitting experimental data to mathematical expressions and doing the spatial projection of the result at local, regional, and continental scales. The usefulness of the platform is illustrated for predicting the potential ecological fitness and spatial variations of the virulence of an entomopathogen fungal-based biopesticide used in an integrated pest management (IPM) context.

The sections below are organized as follows: Section 2 provides details of the methodology starting with the input datasets, followed by the modeling framework, and ending with the architecture of the DSS. Section 3 presents a case study where we demonstrate the application of the DSS. Section 4 discusses the results, and Section 5 provides the conclusion with areas of potential future investigation.

2. Materials and Methods

2.1. Input Dataset

To use the platform for fitting and mapping operations, some input data are required and should be collected and organized in specific formats, as described below.

2.1.1. Experimental Data

The fitting process takes, as input, experimental data recorded from observations from the field or laboratory. In the context of biological control used as the case study here, the virulence data have to be obtained from laboratory experiments replicated over a range of different temperatures under which both the insect and the entomopathogen can interact. Data recorded from the experiments are structured and organized per replicate and temperature, then used as inputs to the platform. Temperature was selected as the key variable due to the paramount role it plays in the development, survival, reproduction, and mortality of Entomopathogenic Fungi (EPF) and insects. A time step of hour, day, or week is necessary for each replication, which allows observing changes in the dependent variable (mortality). Records used as input into the platform are structured and organized as presented in Figure A1 (Appendix A).

2.1.2. Climate Data

Climate data are useful during the mapping process and the spatial projection of the model. Data such as temperature of the studied area are extracted from global climate data (http://www.worldclim.org/). These climate layers contain monthly minimum, maximum, and mean temperatures organized in raster files with "Flat" format (*.flt* and *.hdr* file), with a spatial resolution ranging from 30 arc-seconds (~1 km) to 10 arc-minutes (~20 km) [19].

2.2. Model Fitting Description

The model fitting process can be divided into three stages: pre-regression, regression, and post-regression steps [20]. The pre-regression stage mainly consists of the selection of a set of mathematical expressions to be fitted to the data. The regression stage consists of selecting the proper method for parameter estimation, while the post-regression stage consists of sets of activities necessary to evaluate the model.

To illustrate the model fitting process, the following mathematical annotation is used:

$$z = f(x, \beta) + \theta,$$

where z is the response variable, x represents the input variables, β is the parameter of the model to be estimated, and θ is the error. f is a mathematical function such as an exponential or logarithmic function (see Table 1 for more examples), whose expression depends on x and β. For example, if f is the exponential function with the mathematical expression $f(x) = b(x - x_b)^2$, you have $\beta_f = \{b, x_b\}$.

The ability to identify and include the right variables in a model is equally important to the type of model chosen. We consider a group of m observations $\{z_i\}$ on variables $\{x_i\}$ and try to estimate parameter β by minimizing θ. The minimization of θ follows the least-square curve-fitting procedure using the Levenberg–Marquardt (LM) algorithm. For a given equation $f(x,\beta)$, parameters are estimated such that the sum of square of the deviations $s(\beta)$ minimizes θ via the following expression:

$$s(\beta) = \sum_{i=1}^{m} \left[z_i - f(x_i|\beta)\right]^2.$$

The LM algorithm combines the gradient descent and the Gauss–Newton methods. The gradient descent method reduces the sum of the squared errors by updating the parameters in the downhill direction (the direction opposite to the gradient of the equation), while the Gauss–Newton method reduces the sum of the squared errors by assuming the least-square function is locally quadratic with parameters near the optimum. The fitting procedure starts with an initial value of x_0; then, x is adjusted by Δ only for downhill steps verifying $(J^T J + \lambda I)\,\Delta = J^T r$, where J is the Jacobian matrix of derivatives of the residuals for the parameters, λ is the damping parameter between the two steps, and r is the residual vector. A set of equations used in the study that can be selected by the user are presented in the Supplementary Materials.

Table 1. Summary of key functions used for fitting. The "Name" column gives the name of the model, the "Equation" column gives the mathematical expression of the model, the "comment" column gives the number of derived sub-models from the original main model, and the last column gives the reference for the model. *T* is the independent variable, and *m(T)* represents the virulence model. ID—identifier.

ID	Model Name	Model Main Mathematical Expression	Comment	Reference
1	**Sharpe and DeMichele**	$m(T) = \frac{p.\frac{T}{T_0}.e^{[\frac{\Delta H_A}{R}(\frac{1}{T_0}-\frac{1}{T})]}}{1+e^{[\frac{\Delta H_L}{R}(\frac{1}{T_L}-\frac{1}{T})]}+e^{[\frac{\Delta H_H}{R}(\frac{1}{T_H}-\frac{1}{T})]}}$	12 sub-models Sharpe and DeMichele 1–13	Sharpe and DeMichele 1977
2	**Deva**	$m(T) = b(T - T_{\min})$ T ≥ Tmax $m(T) = 0$ T < Tmax	1 sub-model Deva 1 and 2	Dallwits and Higgins 1992
3	**Logan**	$m(T) = Y * (e^{p*T} - e^{(p*T_{\max} - \frac{(T_{\max}-T)}{v})})$	4 sub-models Logan 1–5	Longan 1976
4	**Briere**	$m(T) = a * T(T - T_o)\left(\sqrt{T_{\max} - T}\right)$	1 sub-model Briere 1 and 2	Briere et al. 1999
5	**Stinner**	$m(T) = \frac{R_{\max}\left(1+e^{k_1+k_2(T_{opc})}\right)}{1+e^{k_1+k_2(T)}}$	3 sub-models Stinner 1–4	Stinner et al. 1974
6	**Hilber and Logan**	$m(T) = Y\left(\frac{T^2}{T^2+d^2} - e^{-\frac{(T_{\max}-T)}{v}}\right)$	2 sub-models Logan 1–3	Hilber and logan 1983
7	**Lactin 1**	$m(T) = e^{p*T} - e^{-\frac{(p*T_l-(T-T_l))}{dt}} + \lambda$	2 sub-models Logan 1–3	Lactin et al. 1995

The selection of mathematical expressions for the fitting process is guided by the type of data available to fit, the knowledge of the system, and its boundaries conditions. The estimation of mathematical expression parameters is directly linked to the convergence of the fitting algorithm and the initial conditions [3].

2.3. Features of the DSS

Spatial visualization emerged as the field that combines the abilities of human perception through data picturing and computer simulation using analytics at the landscape level. Spatial visualization was successfully applied in various studies, and the current platform is inspired by these concepts.

The main features of the platform and how they are linked is described in the unified modeling language (UML) use case diagram shown in Figure 1.

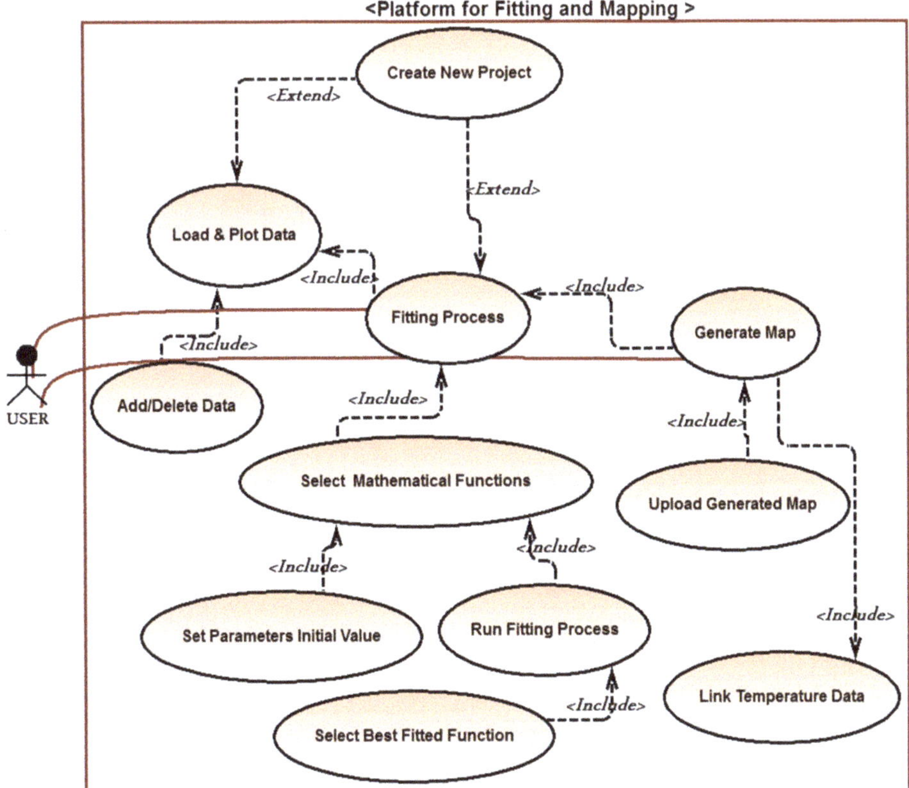

Figure 1. Use case diagram presenting the interactions among the platform components and their functionalities. Each circle represents a feature implemented within the decision support system (DSS). The links from the user to a feature represent the direct interactions of the user with the DSS, while the arrows with labels "include" characterize the relationships and level of dependency between features.

The model fitting and mapping process are done through an interactive process that involves the user in all steps to ensure good and reliable decisions are taken, based on graphical outputs and statistical criteria available in the DSS. The process is summarized in Figure 2.

The fitting procedure begins with a preliminary visual display of input data using statistical aggregation functions to provide the average value for all replicates of the experiment. Then, among the 82 nonlinear mathematical expressions, those which can better fit the data are selected. The database of mathematical expressions is obtained from the literature, and each equation is selected based on its ability to capture the relationship that exists between the dependent and the independent variable. The functions used during this process are implemented in R computer language. The implementation of the Levenberg–Marquardt algorithm in R-package minpack.lm is used for fitting the mathematical expressions to data and estimating the parameters [21–23]. R-squared, R-squared adjusted, Akaike information criterion (AIC), root-mean-squared error (RMSE). and the sum of squared residual are proposed to help to choose the best-fitted model. Once the fitting process is completed, the information is saved and transferred to the mapping perspective.

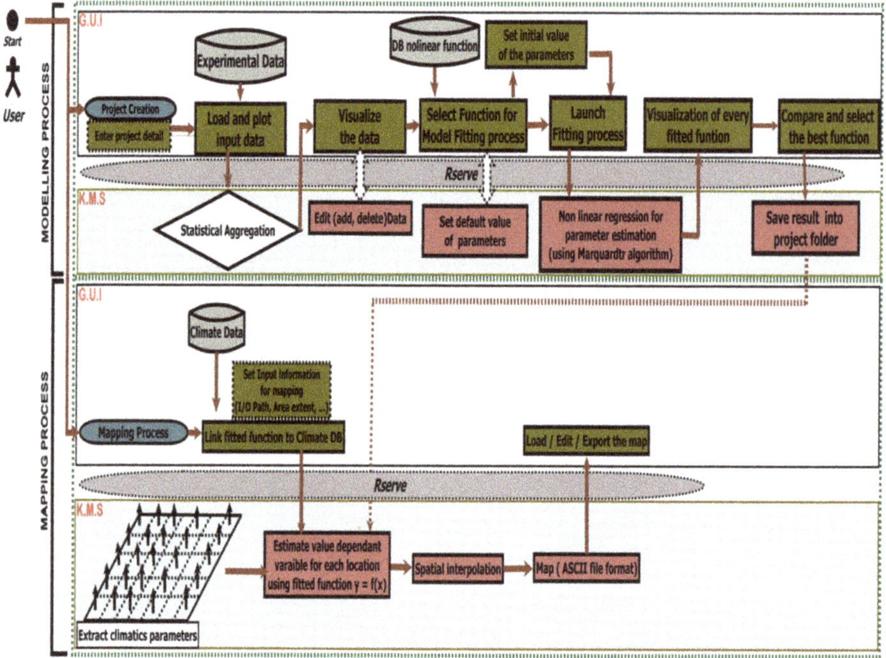

Figure 2. Detailed flowchart diagram of the platform; the figure displays two processes: the fitting process in which experimental data are fitted to nonlinear mathematical expressions using the Marquardt optimization algorithm and the mapping process that begins by linking the obtained best-fitted model to the climate database for map creation. The GUI represents the graphical user interface, and the KMS is the knowledge management system that processes all the simulations.

The mapping exercise requires spatial data as input, which are organized by grids that allow applying the model at the landscape level. From the graphical user interface (GUI), a request is made to a specific studied region, and the values of temperature in the selected locations are extracted from the database in the form of monthly and annual averages. A point object is used to pick the model mathematical expression, and it is consecutively applied in each geographical coordinate of the region of interest to estimate the value of the dependent variable. This allows the value of the dependent variable at each location to be estimated and stored in an ASCII file format (.asc). Additionally, the mapping perspective includes some Udig features for map editing and viewing. It further offers the possibility to transfer the ASCII file obtained to any geographic information system (GIS) software for processing (see Figure A2, Appendix A, for a summary of the algorithm).

2.4. Software Design and Architecture

The tool is designed using the component-based software architectural approach centered on the Rich Client Platform (RCP) framework. The adopted approach for the software development is proven to simplify and facilitate the productivity and the quality of the end product by enhancing the concepts of software reuse, modularity, extensibility, customization, and reduction of development time [24]. The platform extends all the properties of the Udig framework (Figure 3). Moreover, software developed using RCP can run on a variety of operating systems (e.g., Windows, Linux, or Mac).

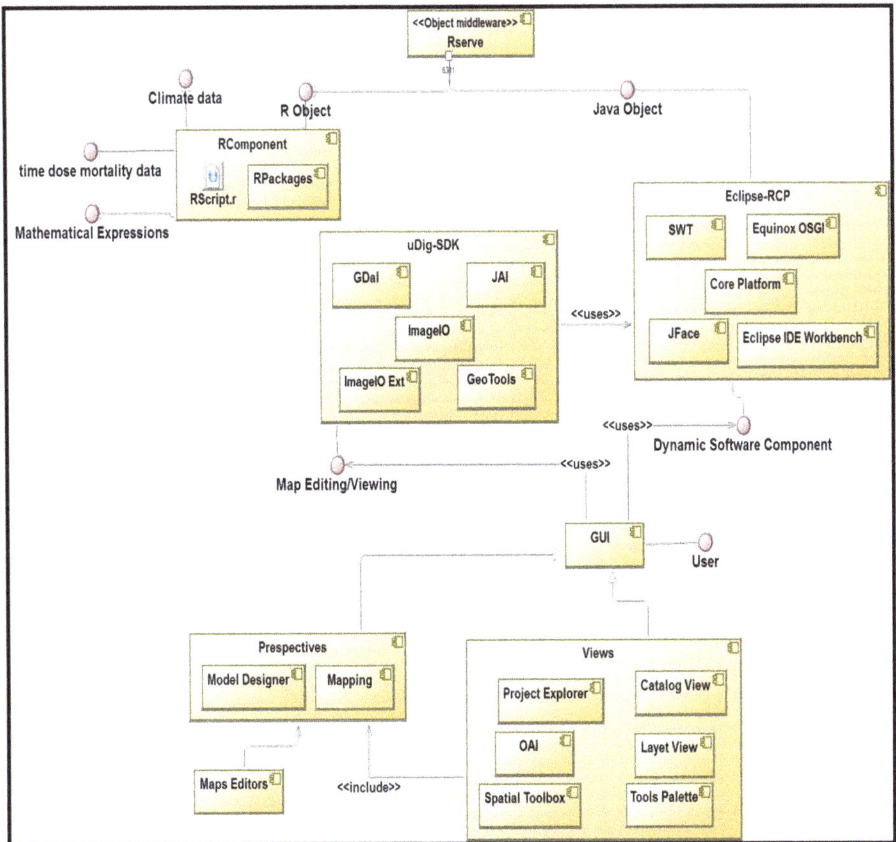

Figure 3. Unified modeling language (UML) component diagram. The users interact with the inbuilt software environment through the perspectives (model designer and mapping perspective) of the GUI. The GUI is based on the Eclipse-RCP (Rich Client Platform) and Udig-RCP components. Rserve allows the communication between R and Java by creating objects with port 6311.

The Eclipse integrated development environment (EDI), which is one of the most commonly used tools for the programming of component-based software [24], was used to develop the tool (see "Supplementary Materials" for the link to source code and setup). Eclipse has the advantage that it allows modular and object-oriented programming, which accepts simple development and extension of the tool product [24]. The platform is structured into five major components (see Sections 2.4.1–2.4.5) including the RComponent, Eclipse RCP, Rserve, Udig-SDK, and the graphical user interface (GUI) (Figure 3). A procedural approach is used in RComponent for the fitting of mathematical equations and statistical criteria, while object-oriented programming is used to interconnect all the components and implement the software GUI.

2.4.1. RComponent

RComponent integrates the features to import data, carry out the analytics, and display the results. RComponent consists of R-scripts encompassing the mathematical expressions, fitting algorithms with procedures for equations and parameter estimations, statistical criteria for selection, and the mapping environment. RComponent enables selecting the equation that best fits the data, which is then applied to temperature raster files for spatial projection. Different R packages (minpack.lm, MASS for nonlinear regression and equation fitting; sp, maptools, rgdal, maps, doRNG for spatial operations leading to the creation of the map) are used within RComponent [25–30]. The *nls.lm()* function implemented in the minpack.lm package was used to simulate the Levenberg–Marquardt algorithm. Spatial objects representing climate data are imported using the function *readGDAL ()* from the rgdal package and *writeAsciiGrid(),* while the maptools package is used to export spatial grid data into ASCII format. Commonly used statistical criteria for comparing and selecting mathematical expressions such as R-squared, R-squared adjusted, Akaike information criterion (AIC), root-mean-squared error ($RMSE$), and the sum of squared residual (SSR) are implemented into RComponent of the platform.

2.4.2. Eclipse RCP

The Eclipse RCP component enables the building of fast and reliable end-user interfaces to guide users during the interactions with the software functionalities. This component consists of different low-level frameworks that offer ways to rapidly develop client-side applications. The software includes the following key features from the RCP framework: the Equinox OSGI (Open Services Gateway initiative) for describing the modular approach used in Java application under Eclipse; the core platform that contains the runtime engine responsible for the run of plug-ins; the JFace dialog preference and wizard framework; the Standard Widget Toolkit (SWT), which provides objects with subclasses of image, color, and font; the Eclipse IDE workbench. All of these features are used to provide a user-friendly and flexible solution.

2.4.3. Rserve

Rserve is a software component used for the creation of objects and the evaluation of the RComponent scripts, as well as their integration into the Java application [31]. This component is used in the developed tool as an object middleware to enable communication and exchange of information between RComponent and the Eclipse RCP component. It is also called an object request broker, which allows the application to send objects and request responses via object-oriented systems. Once a request is issued, Rserve opens a connection to collect the parameters of the request and establishes another connection with RComponent to provide resources for the execution of the request. When the task is completely executed, the output produced by RComponent is sent back to the GUI.

2.4.4. Udig-SDK

Udig-SDK is an open-source desktop application framework built with Eclipse RCP technology, which can be used as a plug-in in RCP applications. To handle the mapping process, the software extends Udig properties to provide a complete Java solution for desktop geographic information system (GIS) data access, editing, and visualization of maps. It is based on GeoTools, ImageIO-Ext, ImageIO, JAI (Java Advanced Imaging), and Gda; the Udig-SDK component allows the mapping of spatial data. Among the key features offered by Udig for map edition, we have layers, style, (AOI (Area of Interest), and catalog. These features allow the user of the developed tool to add or create layers and customize maps.

2.4.5. Graphical User Interface (GUI)

This component is the direct link between the user and the software system. Different perspectives are used to bring key functionalities together in a screen layout that is simple and interactive. The software GUI consists of two perspectives: the model designer and the mapping perspective. The model designer displays functionalities needed for the analytics to (i) import, modify, plot, and visualize experimental data, and (ii) carry out equation fitting. The mapping perspective includes all functionalities for the mapping process. Both perspectives include views, a toolbar, a menu bar, and other graphic components needed for a user graphic interface.

2.4.6. DSS Output Evaluation

The developed DSS produces two main outputs: the model that best fit the experimental data and the geographical distribution map displaying areas of suitability or application of control measures. To evaluate the prediction accuracy of the outputs, users are requested to conduct additional investigations in natural conditions and use the model (if fitted with data from the laboratory) to mimic the field behavior or vice versa. When the output of the DSS is a map, ground scouting is necessary to confirm the projections.

3. Case Study: Using the DSS to Fit Time–Dose–Mortality Data to Mathematical Expressions and Mapping the Potential Zone of Efficacy of Fungal-Based Biopesticides in the Killing of Insect Pests

The case study consisted of using the DSS features to map the potential zone of efficacy for the virulence of the biopesticide *Metarhizium anisopliae* isolate ICIPE 62 against mustard aphids.

Fungi are the most widespread entomopathogenic organisms in terrestrial ecosystems [32,33], gaining many interest for their potential use as biopesticides, due to their ability to infect and kill insects [32,34,35]. Their efficacies were demonstrated both in laboratories and in the field, leading to a worldwide increasing interest in the development of EPF-derived products for commercialization. They are, however, highly influenced by biotic (e.g., pest dynamics) and abiotic (e.g., sunlight, rainfall, temperature, humidity) factors. The platform is used to fit the time- and temperature-dependent mortality data to temperature-dependent virulence mathematical expressions. Thereafter, the virulence equation that best fitted the data is spatially projected to map the areas of the potential efficacy of the biopesticide in controlling the targeted pest.

Data on the mortality (dependent parameter) of mustard aphids caused by the biopesticide *Metarhizium anisopliae* isolate ICIPE 62 were obtained in the laboratory at five different temperatures (independent parameter): 10, 15, 20, 25, and 30 °C. Many isolates of *Metarhizium anisopliae* were reported to have high pathogenicity against several insect pests [36–38]. *Lipaphis pseudobrassicae*, commonly known as turnip aphid or mustard aphid, is a pest that can feed on many types of crops; it was recorded from host plants belonging to over 10 families including Brassicaceae, Cucurbitaceae, and Solanaceae [39]. This pest is globally distributed with the highest level of infestation occurring in Africa, while it is also present in Europe, Asia, and America [40]. The optimal temperature for its development is reported to be around 20 °C [41]. The virulence map was produced for Kenya and Cameroon after completing the fitting process. The spatial evaluation of the generated map was made by comparing the known locations of successful field application of the biopesticide in controlling targeted insect pests to the predicted level of virulence. Illustrations are done below in an interactive way through the features of the designed tool.

3.1. Data Input, Visualization, and Model Fitting Features

The fitting exercise starts with the creation of a new project using the provision of general information such as the project name and name of the EPF, the name of the author, and a brief description of the project. After the creation of the project, the default display in the environment of the tool is the model designer perspective, which is interactive and intuitive in guiding the user through different steps. The independent and dependent variables, in this case, are temperature and insect mortality due to the interaction with EPF, respectively. For example, the virulence rates of the EPF on the targeted insects at temperatures of 15 and 25 °C are 0.7 and 0.89, respectively (Figure 4). The next step consists of loading the experimental data file, which is displayed in Figure 5 on the left side of the user frame. On the right side of the GUI, the recorded mortality is plotted against the corresponding temperature values (Figure 4). Furthermore, the tool offers the user the possibility to include additional temperature values not included in the initial input data file. The new data could be obtained from the literature.

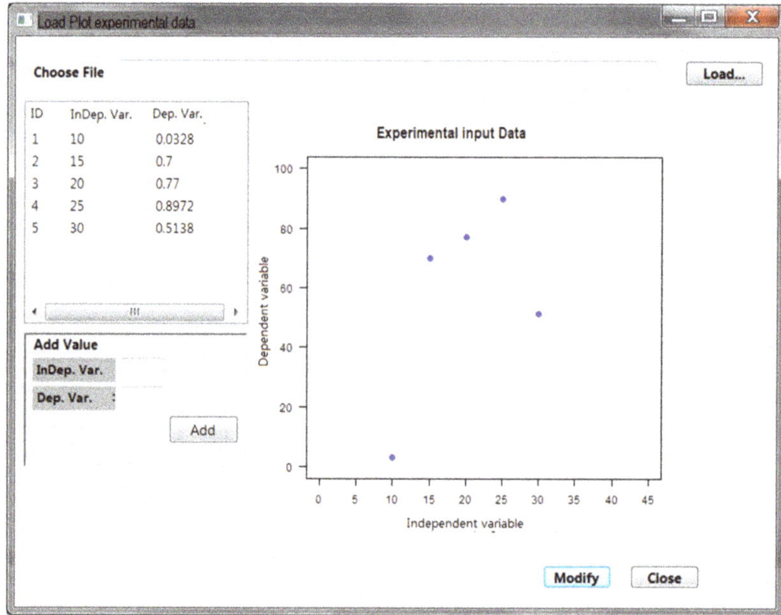

Figure 4. User interface (UI) for importing, plotting, and modifying experimental data.

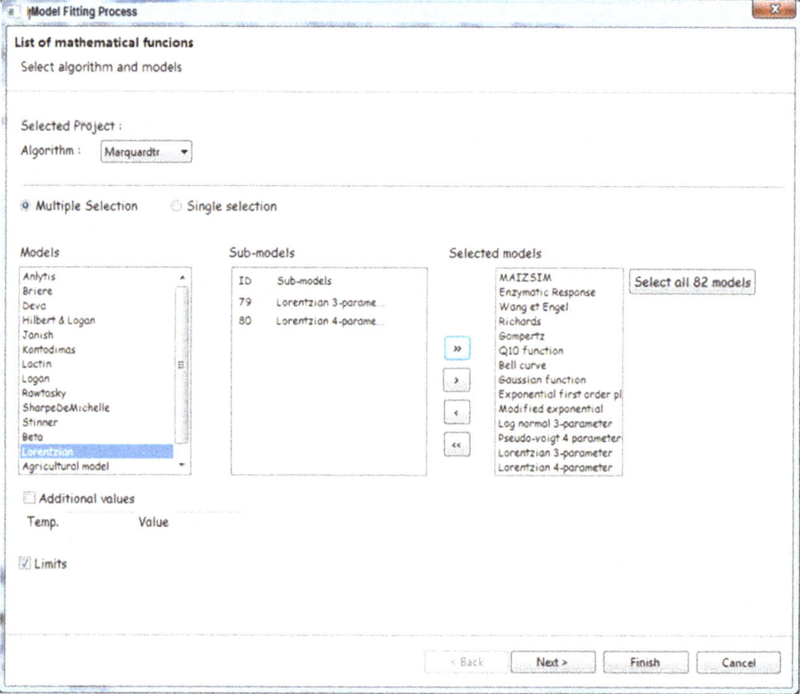

Figure 5. Wizard for the selection list for the fitting process. This frame assists the user in the selection of equations to be fitted with experimental data.

The user can choose to fit a single or many mathematical equations among the 82 available in the software to the input data (Figure 5).

For each equation selected to be fitted to the data, the graph and statistical parameters are generated and displayed in the user interface (Figure 6). By combining the graphical display of the output with the different values of the goodness of fit (Figure 7), the fitted equation with the best performance is selected as the temperature-dependent virulence model for mapping purposes.

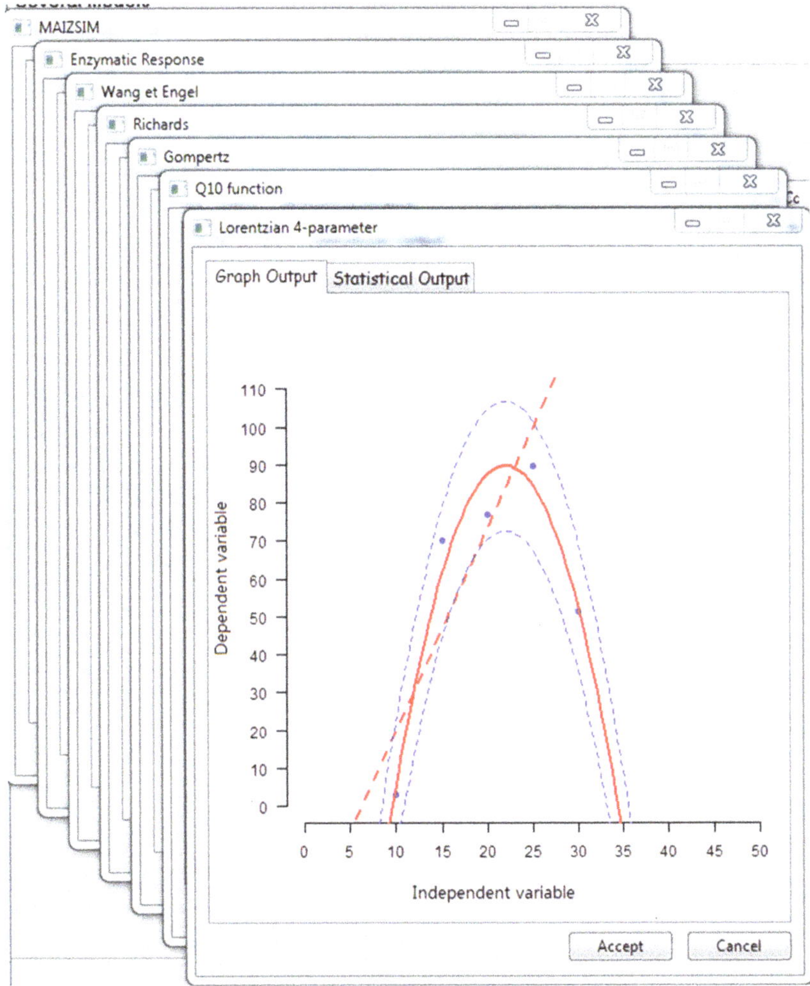

Figure 6. Display of all the fitted equations for the selection of the model.

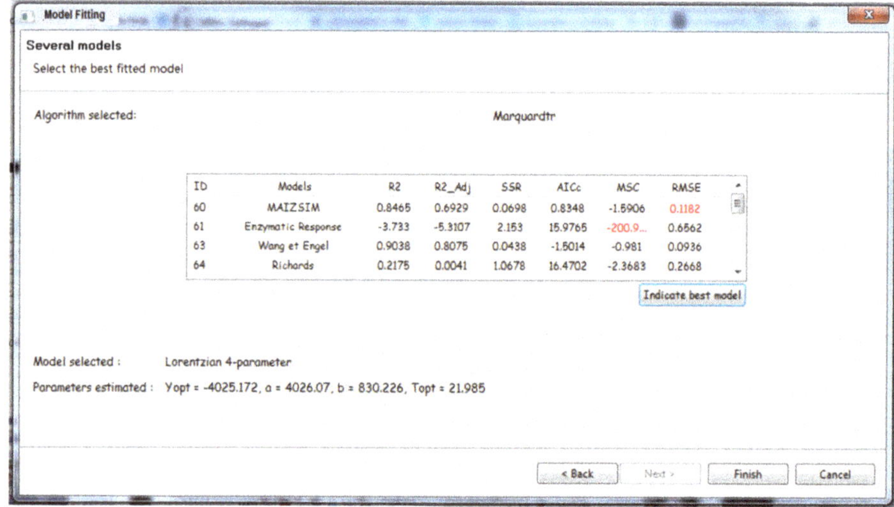

Figure 7. Evaluation criteria and goodness of fit for each fitted equation. Cells in red suggest the best-performing function for each evaluation criterion.

3.2. Mapping Features

Within the mapping perspective, the selected model is linked to climatic data to generate the map (ASCII file) of the potential virulence efficacy. Layers for minimum and maximum temperatures can be imported, and the extent of the area of interest can be defined. Optionally, a filter can be applied to limit the minimum and maximum temperature values that are considered suitable for the process. After the model is linked with a climatic database, a map is produced through a spatial interpolation technique. The user has the option to use the features provided through Udig for additional editing and layouts of the efficacy map or to transfer the generated ASCII file to another GIS software such as QGIS.

The results considered in the case study indicate that the optimum temperature to apply the EPF isolate ICIPE 62 with the highest level of virulence is 22 °C (T_{opt} = 21.98 °C ± 0.29). Based on the Lorentzian four-parameter model results, the maps of the potential zones of the efficacy of EPF isolate ICIPE 62 when applied against mustard aphid were produced for Kenya and Cameroon (Figures 8 and 9, respectively). A level of efficacy that varies between 0 and 1 characterizes the virulence level of ICIPE 62 against the targeted pest.

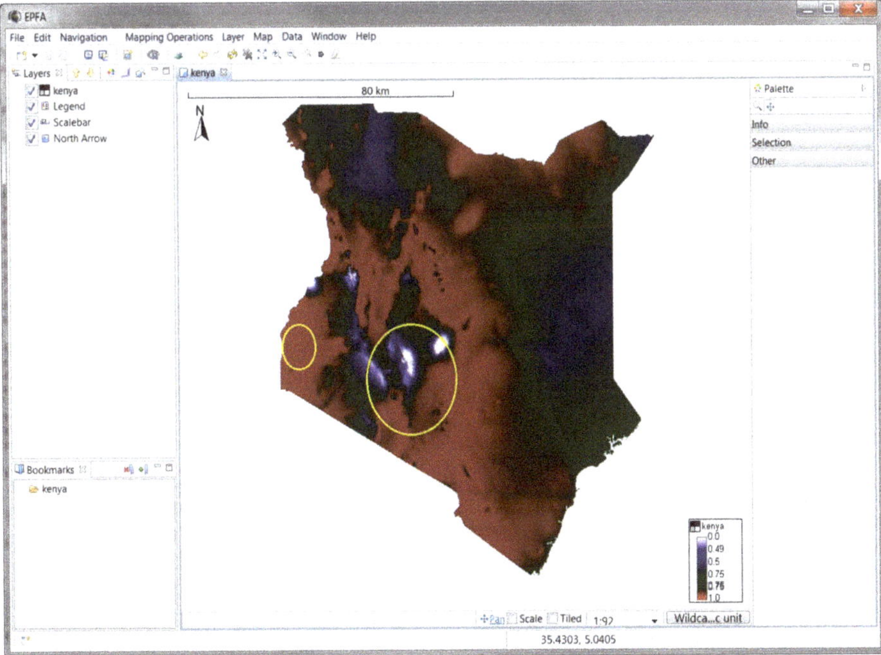

Figure 8. Kenyan map of the potential efficacy of ICIPE 62 isolate when used against mustard aphid, modeled with the software. The level of efficacy varies between 0 and 1. Locations with 0% level of efficacy are displayed in white, values between 0 and 0.5 are displayed in blue, values between 0.5 and 0.75 are displayed in green, and values between 0.75 and 1 are displayed in red. Red indicates the highest efficacy levels. The yellow circles surround areas in Kenya where the EPF isolate ICIPE 62 is successfully used, and results were used for validation of the developed model.

Locations with a virulence level of efficacy ranging from 0 to 0.5 are locations with an average environmental temperature below 15 °C. Locations with a virulence level of efficacy ranging from 0.5 to 0.8 are locations with an average environmental temperature varying from 15 °C to 17 °C and from 26 °C to 30 °C. Locations with a virulence level of efficacy ranging from 0.8 to 1 correspond to locations with an average environmental temperature varying from 17 °C to 26 °C.

To evaluate the prediction accuracy of the output map, field releases of the EFP were conducted at the point locations circled in Figure 8. The outcome, which was a successful control measure, is compared with the model-predicted level of virulence that is greater than or equal to 50%.

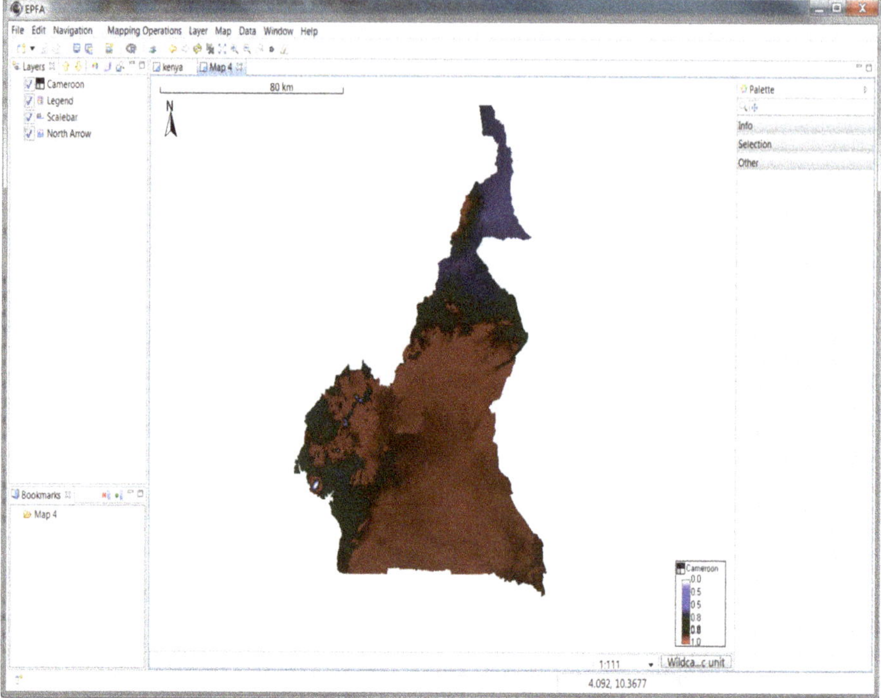

Figure 9. Cameroon map of the potential efficacy of ICIPE 62 isolate when used against mustard aphid, modeled with the software. The level of efficacy varies between 0 and 1. Locations with 0% level of efficacy are displayed in white, values between 0 and 0.5 are displayed in blue, values between 0.5 and 0.75 are displayed in green, and values between 0.75 and 1 are displayed in red. Red indicates the highest efficacy levels.

4. Discussion

This paper presented the design and implementation of an interactive platform for fitting experimental data to mathematical expressions and generating the spatial projection of the result at the local, regional, and continental scales. The platform was developed with the combination of R programming language, RCP architecture, and Udig, which makes it an interactive decision support tool with multiple components, as described in the literature [12–14]. The most important components of the platform are RComponent and the graphical user interface (GUI). However, the trigger of the iterative procedure is the acquisition of the experimental data used as input.

The fitting procedure implemented in the platform applied the LM algorithm. The LM algorithm combines two approaches; it operates like the gradient descent method when the equation parameters are far from their optimal values, while it performs like the Gauss–Newton method when the parameter values are close to their optimum (Nelles, 2001). The main advantage of the LM algorithm is its ability to converge to optimum values faster than the gradient descent or the Gauss–Newton method. This allows the LM algorithm to still find the optimal value of parameters when the initial values of equation parameters are distant from the optimum. This makes the LM algorithm very robust when compared to other minimization algorithms usually used in the fitting procedure (Nelles, 2001).

The GUI is considered as a key component of the tool as it is used directly by the end-user to interact and communicate with the system. The design of the software GUI satisfies the requirements for a software development interface as described in Reference [42]. The component-based approach is adopted to rely on the possibilities to extend and reuse developed components. By employing such a development concept, the maintenance of the software architecture can be secured while new functionalities can be easily integrated. Component-based UIs further enable functional and logical decomposition of the software GUI into perspectives, which helps in defining features needed by the end-users. Component-based UIs accelerate the development process, which is contrary to the iterative or agile development approach that tends to slow the process of developing software. The modularity offered by the RCP component allows easily integrating additional features in the tool.

In comparison to other fields such as education or finance, the need for such tools in agriculture and IPM research activities is lacking despite their importance. This is particularly required to assist in monitoring numerous processes in the crop production system, such as field release of biological control agents. In recent years, progress was perceptible with the proliferation of DSS [43,44], which can be web-based [43] or standalone like the proposed DSS. Despite disparities among conceptual approaches, what they all have in common is the emphases put on input data and the user interface. One advantage of a web-based DSS is the possibility it offers to directly access real-time environmental data to carry out analytics [44], which make these categories of tools highly dependent on an internet connection that is not always available in most of Africa. A standalone DSS, once installed, only requires acquiring input data to be used anywhere and offline. Moreover, the uniqueness of this tool is the flexibility it offers to easily provide spatial distribution maps using gridded values of temperature for delimiting potential areas of field application of control measures. A similar attempt was presented in References [45,46]; however, it is not directly applicable as users still need to be knowledgeable in statistical and geospatial sciences. The current platform generalizes and combines all the steps used in these studies to provide researchers, with no computational skills, the opportunity to process the fitting and select the models that can further be projected spatially at the landscape level.

Application: The Lorentzian 4-parameter model obtained in the application section estimates an optimum temperature for the higher virulence of ICIPE62 in killing the aphid at about 21 °C. When comparing the current distribution map of the *Lipaphis pseudobrassicae* [40] with the map of the potential zone of the efficacy of ICIPE 62 isolate in Kenya and Cameroon, we observed that many invaded locations fit well with a potential zone of efficacy with virulence level greater than or equal to 0.5. Although the estimates were made without full inclusion of other environmental variables that have impacts on the fungi efficacy in killing insects, the outputs are promising. However, it will be useful to explore the association of temperature with other factors such as relative humidity to improve the accuracy of the prediction, especially for mapping the virulence level of the EPF. Indeed, many studies highlighted the key role played by both temperature and relative humidity in enhancing the virulence of EPF on insect pests [47–49]. On this note, a good perspective to consider for improving the current tool will be to consider the association of at least two factors (temperature and relative humidity, for example) as independent variables in the fitting process.

5. Conclusions and Future Works

Herein, we presented an interactive generic platform for predicting agro-ecological processes through the use of experimental data that are fitted to mathematical expressions, whereby the resulting best-fitted equation can be applied at the landscape level. The platform is a useful tool for anyone interested in fitting and conducting a spatial visualization of data. The current platform could be of great help for science pathologists and IPM practitioners worldwide in their attempt to increase the use and application of control measures within an agriculture IPM context. To further improve DSS development, the economic injury level (EIL) concepts that rely on economic threshold models may be added as a guide in evaluating the cost and benefit of deploying any control measure. The added module will integrate a combination of environmental and biological factors which are linked to economic factors, yielding an environmental economic threshold module. However, to avoid complexity, it is also possible that outputs from the developed tool can be used as input in other software specifically developed to conduct econometric analysis.

Supplementary Materials: The source code of the software is available online at https://github.com/Atoundem/EPFA_RCP. The setup and installation requirements can be downloaded at https://github.com/Atoundem/EPFA_RCP/tree/master/Setup_Intall_Requirement.

Author Contributions: Conceptualization, S.E. and H.E.T.; Methodology, R.A.G. and H.E.Z.T.; Software, R.A.G. and H.E.Z.T.; Writing—original draft preparation, R.A.G.; Writing—review and editing, R.A.G., L.B.-F., C.B. and H.E.Z.T.; Supervision, S.A.M., W.M. and H.E.Z.T.; Funding acquisition, S.E. and H.E.Z.T. All authors have read and agreed to the published version of the manuscript.

Funding: This work was supported by the Volkswagen Foundation under Grant [VW-94362].

Acknowledgments: The first author of this study is a PhD student working in the fellowship project (VW-89362) of Henri E.Z. Tonnang funded by the Volkswagen Foundation under the funding initiative Knowledge for Tomorrow: Cooperative Research Projects in Sub-Saharan on Resources, their Dynamics, and Sustainability—Capacity Development in Comparative and Integrated Approaches. The authors thank the Federal Ministry for Economic Cooperation and Development (BMZ), Germany, which provided the financial support through the Tuta IPM project, the German Academic Exchange Service (DAAD), and the STRIVE project funded by the German Federal Ministry of Education and Research.

Conflicts of Interest: The authors declare no conflict of interest.

Appendix A

Var1	Var2	Var3	Var4	Var5
15	3	2	20	0
15	3	3	20	2
15	3	4	18	0
15	3	5	18	2
15	3	6	16	2
15	3	7	14	0
20	1	1	20	0
20	1	2	15	5
20	1	3	12	3
20	1	4	10	2
20	1	5	8	2
20	1	6	6	2
20	1	7	6	0
20	2	1	20	0
20	2	2	16	4
20	2	3	12	4
20	2	4	8	4
20	2	5	7	1
20	2	6	7	0
20	2	7	7	0
20	3	1	18	2
20	3	2	16	2
20	3	3	10	6
20	3	4	7	3
20	3	5	7	0
20	3	6	7	0
20	3	7	7	0
25	1	1	20	2
25	1	2	18	0
25	1	3	15	3
25	1	4	10	5
25	1	5	8	2
25	1	6	5	3
[illegible]	[illegible]	[illegible]	[illegible]	[illegible]

Figure A1. Experimental data: Var1 represents the range of the independent variable (temperature); Var2 is the number of replicates in the experiment, Var3 is the time duration of each experiment; Var4 and Var5 are records of the dependent variable (mortality) with the variation of the independent variable.

Algorithm: *model fitting and mapping*

Input Variables:

ListFunctionParameters. $\beta = (\beta_1, \beta_2, \ldots, \beta_m)$ // β_i list of parameters for the function Z_i

experimentalData: Represent the data collected form complex process to be fitted.

climateDatabase: Represent the area targeted to the mapping process.

Begin:

Co: **Fitting process** Fco

Initialization $Z_i = f(x_i, \beta_i) + \theta_i$

Import all required packages

$\beta \leftarrow$ set parameters initial value $(\beta_1^0, \beta_2^0, \ldots, \beta_m^0)$

$Z = (Z_1, Z_2, \ldots, Z_m) \leftarrow (\beta_1^0, \beta_2^0, \ldots, \beta_m^0)$ // Initialize each candidate function with parameters

for each Z_i in ***ListCandidateFunction***

$Z_i = f(x_i, \beta_i) + \theta_i$

Plot (***experimentalData***) // Visualize the trend of experimental data

Plot (Z_i, β_i^0) // visualize the trend of the function with initial parameter value

$\beta_i \leftarrow$ nls.lm(par $=\beta_i^0$, ..., jcall = j, x = x_i, y = Z_i) // Nonlinear Reg. for parameter estimation

$\theta_i \leftarrow$ Estimated model and parameters errors

Compute model Evaluation criteria (R2, R2_Adj, AIC, MSC, RMSE)

Plot (Z_i, β_i, θ_i) // Visualize the output of fitted function

EndFor

m(x) $\leftarrow$ selectedBestModel ($\{Z_i\}$) // Selection of the best model base on evaluation criteria

Co **End fitting process and beginning of mapping process** Fco

(Lat_{min}, Lat_{max}, $Long_{min}$, $Long_{max}$) $\leftarrow$ read (***climateDatabase***) // Import climate database

MatLocation $\leftarrow$ Build the matrix of georeferenced location within (Lat_{min}, Lat_{max}, $Long_{min}$, $Long_{max}$)

for every (lat_j, $long_k$) in ***MatLocation***

$Param_{(j,k)} \leftarrow$ Extraction (lat_j, $long_k$) // Extract value of model independent variables

$Mat_{output} \leftarrow$ Estimate ***m(*** $Param_{(j,k)}$) // estimate dependent variables using fitted model

EndFor

Map.asc $\leftarrow$ SpatialInterpolation (Mat_{output}) // generate the map with spatial interpolation.

AEZ (Map.asc) // Map spatial classification through Agro-Ecological Zoning process

Plot (Map.asc) // Visualize the generated map.

Output.

Return: Map of the potential zone of successful implementation of the process

End

Figure A2. Algorithm: model fitting and mapping processes.

References

1. Jones, J.W.; Antle, J.M.; Basso, B.; Boote, K.J.; Conant, R.T.; Foster, I.; Godfray, H.C.J.; Herrero, M.; Howitt, R.E.; Janssen, S.; et al. Brief history of agricultural systems modeling. *Agric. Syst.* **2017**, *155*, 240–254. [CrossRef] [PubMed]
2. Lepenioti, K.; Bousdekis, A.; Apostolou, D.; Mentzas, G. Prescriptive analytics: Literature review and research challenges. *Int. J. Inf. Manag.* **2020**, *50*, 57–70. [CrossRef]
3. Archontoulis, S.V.; Miguez, F.E. Nonlinear Regression Models and Applications in Agricultural Research. *Agron. J.* **2015**, *107*, 786. [CrossRef]
4. Klosterman, R.E. Simple and Complex Models. *Environ. Plan. B Plan. Des.* **2012**, *39*, 1–6. [CrossRef]
5. Eva, Y.-H.; Wu, M.-C. Hung, Comparison of Spatial Interpolation Techniques Using Visualization and Quantitative Assessment. *Appl. Spat. Stat.* **2016**, *11*, 17–34. [CrossRef]

6. Patel, N.R.; Mandal, U.K.; Pande, L.M. Agro-ecological Zoning System-A Remote Sensing and GIS Perspective Upscaling of photosynthesis through Sun-induced fluorescence (SIF) View project 1. Regional Carbon Cycle Modeling for India and surrounding oceans View project. *J. Agrometeorol.* **2000**, *2*, 1–13. Available online: https://www.researchgate.net/publication/270683979 (accessed on 3 April 2020).
7. Gimond, M. Intro to GIS and Spatial Analysis. 2019. Available online: https://mgimond.github.io/Spatial/index.html (accessed on 9 October 2019).
8. FAO. Agro-Ecological Zoning: Guidelines, Rome. 1996. Available online: https://books.google.com/books?hl=fr&lr=&id=IWFD2zGLyrYC&oi=fnd&pg=PA1&dq=AGRO-ECOLOGICAL+ZONING+Guidelines&ots=bAH-Or-Nn0&sig=XdhPDm3WBbN8ckFjP2d5Zj5qaIc (accessed on 3 April 2020).
9. Mardani, A.; Jusoh, A.; Nor, K.M.D.; Khalifah, Z.; Zakwan, N.; Valipour, A. Multiple criteria decision-making techniques and their applications—A review of the literature from 2000 to 2014. *Econ. Res. Istraž.* **2015**, *28*, 516–571. [CrossRef]
10. Belton, V.; Stewart, T.J. *Multiple Criteria Decision Analysis: An Integrated Approach*; Springer: New York, NY, USA, 2002.
11. Stojčić, M.; Zavadskas, E.; Pamučar, D.; Stević, Ž.; Mardani, A. Application of MCDM Methods in Sustainability Engineering: A Literature Review 2008–2018. *Symmetry (Basel)* **2019**, *11*, 350. [CrossRef]
12. Sprague, R.H.; Carlson, E. *Building Effective Decision Support Systems*; Prentice Hall College Div: Englewood Cliffs, NJ, USA, 1982.
13. Dan, P. Ask Dan about DSS—What Are the Components of A Decision Support System? *2005.* Available online: http://dssresources.com/faq/index.php?action=artikel&id=101 (accessed on 5 April 2017).
14. Karacapilidis, N. *An Overview of Future Challenges of Decision Support Technologies*; Springer: London, UK, 2006; pp. 385–399. [CrossRef]
15. Huber, G.P. Organizational science contributions to the design of decision support systems. 1980. Available online: http://pure.iiasa.ac.at/id/eprint/1221/1/XB-80-512.pdf#page=55 (accessed on 17 January 2020).
16. Fick, G.; Sprague, R.H. *Decision Support Systems: Issues and Challenges: Proceedings of the an International Task Force Meeting June 23–25, 1980*; Elsevier: Oxford, UK, 1980. Available online: http://www.sciencedirect.com/science/book/9780080273211 (accessed on 5 April 2017).
17. Wierzbicki, A.P.; Makowski, M.; Wessels, J. Model-Based Decision Support Methodology with Environmental Applications. *Interfaces* **2000**, *32*(2), 84. [CrossRef]
18. Power, D.J. Understanding Data-Driven Decision Support Systems. *Inf. Syst. Manag.* **2008**, *25*, 149–154. [CrossRef]
19. Hijmans, R.J.; Cameron, S.E.; Parra, J.L.; Jones, P.G.; Jarvis, A. Very high resolution interpolated climate surfaces for global land areas. *Int. J. Climatol.* **2005**, *25*, 1965–1978. [CrossRef]
20. Jaqaman, K.; Danuser, G. Linking data to models: Data regression. *Nat. Rev. Mol. Cell Biol.* **2006**, *7*, 813–819. [CrossRef] [PubMed]
21. Marquardt, D. An Algorithm for Least-Squares Estimation of Nonlinear Parameters. *J. Soc. Ind. Appl. Math.* **1963**, *11*, 431–441. [CrossRef]
22. Gavin, H. The Levenberg-Marquardt method for nonlinear least squares curve-fitting problems. 2011. Available online: http://people.duke.edu/~{}hpgavin/ce281/lm.pdf (accessed on 5 December 2019).
23. Lourakis, M.I.A. A Brief Description of the Levenberg-Marquardt Algorithm Implemened by levmar. *Found. Res. Technol.* **2005**, *4*, 1–6. [CrossRef]
24. Silva, V. *Practical Eclipse Rich Client Platform Projects*; Apress: New York, NY, USA, 2009. [CrossRef]
25. Gaujoux, R. doRNG: Generic Reproducible Parallel Backend for "Foreach" Loops. 2017. Available online: https://cran.r-project.org/web/packages/doRNG/index.html (accessed on 9 October 2017).
26. Pebesma, E.; Bivand, R.; Rowlingson, B.; Gomez-Rubio, V.; Hijmans, R.; Sumner, M.; MacQueen, D.; Lemon, J.; O'Brien, J. sp: Classes and Methods for Spatial Data. 2017. Available online: https://cran.r-project.org/web/packages/sp/index.html (accessed on 9 October 2017).
27. Ripley, B.; Venables, B.; Bates, D.M.; Hornik, K.; Gebhardt, A.; Firth, D. MASS: Support Functions and Datasets for Venables and Ripley's MASS. 2017. Available online: https://cran.r-project.org/web/packages/MASS/index.html (accessed on 9 October 2017).
28. Elzhov, T.V.; Mullen, K.M.; Spiess, A.-N.; Bolker, B. Minpack.lm: R Interface to the Levenberg-Marquardt Nonlinear Least-Squares Algorithm Found in MINPACK, Plus Support for Bounds. 2016. Available online: https://cran.r-project.org/web/packages/minpack.lm/index.html (accessed on 9 October 2017).

29. Bivand, R.; Lewin-Koh, N.; Pebesma, E.; Archer, E.; Baddeley, A.; Bearman, N.; Bibiko, H.-J.; Brey, S.; Callahan, J.; Carrillo, G.; et al. Maptools: Tools for Reading and Handling Spatial Objects. 2017. Available online: https://cran.r-project.org/web/packages/maptools/index.html (accessed on 9 October 2017).
30. Becker, R.A.; Wilks, A.R.; Brownrigg, R.; Minka, T.P.; Deckmyn, A. Maps: Draw Geographical Maps. 2017. Available online: https://cran.r-project.org/web/packages/maps/index.html (accessed on 9 October 2017).
31. Urbanek, S. Rserve—A Fast Way to Provide R Functionality to Applications. 2003. Available online: https://www.r-project.org/conferences/DSC-2003/Proceedings/Urbanek.pdf (accessed on 14 April 2017).
32. Augustyniuk-Kram, A.; Kram, K.J. Entomopathogenic Fungi as an Important Natural Regulator of Insect Outbreaks in Forests (Review). 2012. Available online: https://www.intechopen.com/books/forest-ecosystems-more-than-just-trees/entomopathogenic-fungi-as-an-important-natural-regulator-of-insect-outbreaks-in-forests-review- (accessed on 21 September 2016).
33. Lacey, L.A.; Grzywacz, D.; Shapiro-Ilan, D.I.; Frutos, R.; Brownbridge, M.; Goettel, M.S. Insect pathogens as biological control agents: Back to the future. *J. Invertebr. Pathol.* **2015**, *132*, 1–41. [CrossRef]
34. Roberts, D.W.; Humber, R.A. Entomogenous Fungi. *Biol. Conidial Fungi* **1981**, *2*(201), 201–236. Available online: http://linkinghub.elsevier.com/retrieve/pii/B9780121795023500145 (accessed on 21 September 2016).
35. Shahid, A.A.; Rao, A.Q.; Bakhsh, A.; Husnain, T. Entomopathogenic fungi as biological controllers: New insights into their virulence and pathogenicity. 2012. Available online: http://agris.fao.org/agris-search/search.do?recordID=RS2012000992 (accessed on 21 September 2016).
36. Bayissa, W.; Ekesi, S.; Mohamed, S.A.; Kaaya, G.P.; Wagacha, J.M.; Hanna, R.; Maniania, N.K. Selection of fungal isolates for virulence against three aphid pest species of crucifers and okra. *J. Pest Sci.* **2004**, *90*, 355–368. [CrossRef]
37. Migiro, L.N.; Maniania, N.K.; Chabi-Olaye, A.; Vandenberg, J. Pathogenicity of Entomopathogenic Fungi Metarhizium anisopliae and Beauveria bassiana (Hypocreales: Clavicipitaceae) Isolates to the Adult Pea Leafminer (Diptera: Agromyzidae) and Prospects of an Autoinoculation Device for Infection in the Field. *Environ. Entomol.* **2010**, *39*, 468–475. [CrossRef]
38. Niassy, S.; Maniania, N.K.; Subramanian, S.; Gitonga, M.L.; Maranga, R.; Obonyo, A.B.; Ekesi, S. Compatibility of Metarhizium anisopliae isolate ICIPE 69 with agrochemicals used in French bean production. *Int. J. Pest Manag.* **2012**, *58*, 131–137. [CrossRef]
39. Akhtar, K.U.S.; Dey, D. Spatial Distribution of Mustard Aphid Lipaphis erysimi (Kaltenbach) Vis-à-vis its Parasitoid, Diaeretiella rapae (M'intosh). *World Appl. Sci.* **2010**, *11*, 284–288.
40. CABI. Mustard Aphid (Lipaphis Erysimi) Plantwise Technical Factsheet, Plantwise Knowledge Bank. 2020. Available online: http://www.plantwise.org/KnowledgeBank/Datasheet.aspx?dsid=30913 (accessed on 20 June 2017).
41. Awaneesh, Mustard Aphid agropedia. 2009. Available online: http://agropedia.iitk.ac.in/node/4578 (accessed on 29 June 2017).
42. Scott, D. 6 Reasons for Component-based UI Development. 2016. Available online: https://www.tandemseven.com/technology/6-reasons-component-based-ui-development/ (accessed on 14 July 2017).
43. Jones, V.P.; Brunner, J.F.; Grove, G.G.; Petit, B.; Tangren, G.V.; Jones, W.E. A web-based decision support system to enhance IPM programs in Washington tree fruit. *Pest Manag. Sci.* **2010**, *66*(6), 587–595. [CrossRef] [PubMed]
44. Damos, P. Modular structure of web-based decision support systems for integrated pest management: A review. *Agron. Sustain. Dev.* **2015**, *35*, 1347–1372. [CrossRef]
45. Klass, J.I.; Blanford, S.; Thomas, M.B. Use of a geographic information system to explore spatial variation in pathogen virulence and the implications for biological control of locusts and grasshoppers. *Agric. For. Entomol.* **2007**, *9*, 201–208. [CrossRef]
46. Klass, J.I.; Blanford, S.; Thomas, M.B. Development of a model for evaluating the effects of environmental temperature and thermal behaviour on biological control of locusts and grasshoppers using pathogens. *Agric. For. Entomol.* **2007**, *9*, 189–199. [CrossRef]
47. Mishra, S.; Kumar, P.; Malik, A. Effect of temperature and humidity on pathogenicity of native Beauveria bassiana isolate against Musca domestica L. *J. Parasit. Dis.* **2015**, *39*, 697–704. [CrossRef] [PubMed]

48. Hsiao, W.-F.; Bidochka, M.J.; Khachatourians, G.G. Effect of temperature and relative humidity on the virulence of the entomopathogenic fungus, Verticillium lecanii, toward the oat-bird berry aphid, Rhopalosiphum padi (Hom., Aphididae). *J. Appl. Entomol.* **1992**, *114*, 484–490. [CrossRef]
49. Athanassiou, C.G.; Kavallieratos, N.G.; Rumbos, C.I.; Kontodimas, D.C. Influence of Temperature and Relative Humidity on the Insecticidal Efficacy of Metarhizium anisopliae against Larvae of Ephestia kuehniella (Lepidoptera: Pyralidae) on Wheat. *J. Insect Sci.* **2017**, *17*, 22. [CrossRef]

Article

An Evaluation Framework and Algorithms for Train Rescheduling

Sai Prashanth Josyula *, Johanna Törnquist Krasemann and Lars Lundberg

Department of Computer Science, Blekinge Institute of Technology, 37141 Karlskrona, Sweden; johanna.tornquist.krasemann@bth.se (J.T.K.); lars.lundberg@bth.se (L.L.)

* Correspondence: sai.prashanth.josyula@bth.se

Received: 26 October 2020; Accepted: 7 December 2020; Published: 11 December 2020

Abstract: In railway traffic systems, whenever disturbances occur, it is important to effectively reschedule trains while optimizing the goals of various stakeholders. Algorithms can provide significant benefits to support the traffic controllers in train rescheduling, if well integrated into the overall traffic management process. In the railway research literature, many algorithms are proposed to tackle different versions of the train rescheduling problem. However, limited research has been performed to assess the capabilities and performance of alternative approaches, with the purpose of identifying their main strengths and weaknesses. Evaluation of train rescheduling algorithms enables practitioners and decision support systems to select a suitable algorithm based on the properties of the type of disturbance scenario in focus. It also guides researchers and algorithm designers in improving the algorithms. In this paper, we (1) propose an evaluation framework for train rescheduling algorithms, (2) present two train rescheduling algorithms: a heuristic and a MILP-based exact algorithm, and (3) conduct an experiment to compare the two multi-objective algorithms using the proposed framework (a proof-of-concept). It is found that the heuristic algorithm is suitable for solving simpler disturbance scenarios since it is quick in producing decent solutions. For complex disturbances wherein multiple trains experience a primary delay due to an infrastructure failure, the exact algorithm is found to be more appropriate.

Keywords: algorithm evaluation; decision support systems; parallel algorithms; multi-objective optimization; train rescheduling

1. Introduction

In railway traffic systems, whenever disturbances occur, it is important to effectively reschedule trains while optimizing the goals of various stakeholders. In real time, rescheduling of trains during a disturbance is typically carried out manually by a dispatcher [1], or a train traffic controller. In this process, deviations from the original plan and conflicts are detected by constantly supervising the status of traffic and infrastructure [1]. A railway traffic management system (TMS) constitutes remote equipment and software tools which can support the dispatchers in managing (or controlling) a network's railway traffic [2]. Today's dispatching is supported in various ways by TMSs, which typically [3]: (i) show the current status of the railway network in the dispatching area, (ii) show the positions of the trains, status of signals and switches in real time, (iii) predict train movements and detect (potential) conflicts. However, when it comes to conflict resolution, only a few of the currently existing railway TMSs are actually able to compute and suggest alternative rescheduling decisions, let alone incorporate advanced rescheduling algorithms [4]. A TMS that incorporates intelligent, flexible and (semi)autonomous train rescheduling algorithms has several benefits [5], e.g., facilitating the incorporation of important information in the decision-making process, enabling a wider and longer planning horizon, and reducing the work load of the dispatchers/traffic controllers, if incorporated well in the traffic management process. As of 2020, several countries (e.g., Sweden and Switzerland) are preparing to deploy new TMSs in railways that integrate, unify and automate a significant part of the traffic management process. A major advantage of such integrated systems is the possibility to increase the level of automation for rescheduling the railway traffic during disturbances [5].

Several algorithms for tackling the train rescheduling problem have been proposed in railway research publications [6]. There is, however, a need to analyze and compare the effectiveness and efficiency of proposed algorithms, to assess the main strengths and limitations. The benefits of comparing the algorithms and the solutions output by them are threefold: (i) enables the practitioners to select an algorithm suitable for the occurring disturbance scenario, (ii) guides the researchers and algorithm designers in improving their algorithms, and (iii) increases future co-operation among researchers and enables exchange of innovative solutions [7]. In this paper, we propose criteria to consider when evaluating a train rescheduling algorithm or comparing it against other algorithms.

The paper is organized as follows. In the next section, we present the train rescheduling problem and the scope of this study, along with some key terminology. In Section 3, we present an overview of related research work and a brief discussion of the main research challenges addressed in this paper, along with the expected research contributions. Section 4 presents the first part of the framework that serves to classify and compare train rescheduling algorithms on a functional/conceptual level. Section 5 presents the second part of the framework that contains a selection of key aspects proposed to be used in systematic performance evaluation and benchmarking of algorithms. Section 6 presents a description of the two rescheduling algorithms that are used to demonstrate the framework's applicability. Section 7 presents the chosen dataset containing the problem instances and the experimental setup. In Section 8, we demonstrate the application of the framework for evaluating and comparing the performance of the two algorithms. In that section, we present the results from the evaluation and discussions based on our analysis. Finally, we present some conclusions and suggest future work in Section 9.

2. Problem Description, Scope, and Terminology

Railway engines and wagons are known as rolling stock. A train denotes both a composition of rolling stock as well as a timetabled service, allowing the transportation of travellers (passenger trains) or goods (freight trains) between stations [8]. Often, the operation of passenger and freight train services is based on preplanned timetables which ensure operational feasibility of the services by respecting the applicable constraints. A disturbance in a railway network is an unexpected event that renders the originally planned timetable infeasible by introducing 'conflicts'. A conflict is considered to be a situation that arises when two trains are scheduled to occupy an infrastructure resource during overlapping time periods in a way such that one or more system constraints are violated. Several actions need to be taken in real time to prevent or resolve conflicts.

Disturbances are relatively small perturbations in the railway system that can be handled by rescheduling only the railway traffic (i.e., the train timetable) [9]. Disturbances are triggered by incidents such as over-crowded platform(s) that possibly lead to unexpectedly long boarding times and minor delays, or e.g., shorter signalling system failures that may cause more significant delays for several trains. Larger incidents caused by e.g., longer signalling system failures require not only train rescheduling, but also rolling stock and crew rescheduling. Such incidents are often referred to as disruptions [9].

Railway timetables are ideally planned with appropriate time margins in order to enable delayed trains to recover from minor delays and to prevent the propagation of delays from one train to another (i.e., knock-on effects). In case of a disturbance that causes a significant delay to one or more trains, conflicts may arise in the original timetable, thus making it operationally infeasible. The identification and resolution of these conflicts, by adjusting the existing timetable, to obtain a feasible timetable in real time, is known as train rescheduling. The aim of train rescheduling is to quickly obtain a revised feasible timetable of sufficient quality [9]. Train rescheduling is also known as train dispatching and *real-time railway traffic rescheduling*.

When rescheduling trains, the two main stakeholders involved in the process are infrastructure managers and railway operators. The railway infrastructure manager (IM) owns the railway network and associated infrastructure [10]. The IM manages and coordinates all the traffic in the network, both freight and passenger, assuring the operational safety and quality of services [10,11]. The IM also maintains and innovates the rail infrastructure [11]. Examples of IMs are Trafikverket (Sweden), Jernbaneverket (Norway), and Infrabel (Belgium). An important role within the IM organization is that of a dispatcher. Typically, a dispatcher is responsible for monitoring and controlling the railway traffic (i.e., train movements) and rescheduling the traffic plan for his/her control area [12]. The dispatcher is often also responsible for ensuring the safety of scheduled maintenance activities in the railway network. A company operating passenger or freight rail services over the railway infrastructure is called a train operating company (TOC). It is also known as a railway operator or a railway undertaking. Examples are SJ, Tågkompaniet, GreenCargo (Sweden), FlyToget and CargoNet (Norway).

The rescheduling tactics used to prevent and resolve conflicts can be broadly categorized into the following two types: (1) *IM tactics* and (2) *IM + TOC tactics*. IM tactics are typically used by the dispatcher to handle disturbances. Such rescheduling tactics can generally be used without consulting the TOCs and they include (i) retiming (i.e., allocating new arrival and departures times to one or more trains), (ii) local rerouting (i.e., allocating alternative tracks to one or more trains) and (iii) reordering (i.e., prioritizing a train over another). $IM + TOCtactics$ are typically deployed to handle disruptions and they require the dispatcher to consult with the affected TOCs. Examples of such tactics are (i) global rerouting (i.e., changing the route of trains), (ii) train cancellations (partially/fully cancelling the affected services) and (iii) short-turning of trains. Typically, the IM + TOC tactics are considered to be major decisions, compared to the IM tactics. The reason is that the effects of an IM + TOC decision spill over to other organizations and stakeholders' operational plans.

The actions performed by a typical train rescheduling algorithm can be broadly categorized into two main tasks: (i) computing alternative rescheduling solutions and (ii) selecting a solution based on the objective(s). The computation of alternative solutions primarily involves employing the different mentioned rescheduling tactics to resolve identified potential conflicts. The decision maker is the person responsible for making the decisions regarding adjustments in the disturbed train timetable that will lead to a rescheduled timetable. The algorithm might also assist the decision maker in the selection of a revised timetable, for example, by presenting an analysis of the computed alternative rescheduling solutions and ranking the solutions based on a selection of qualitative and quantitative solution quality indicators.

Table 1 gives an example of how a train rescheduling algorithm can assist the decision-making process. The framework proposed in this paper primarily focuses on the capabilities and performance of algorithms, while the interaction between rescheduling algorithms, involved human decision makers, and the traffic management system, is not described in detail.

Table 1. Examples of how train rescheduling can be viewed with/without an algorithm's assistance.

Rescheduling Performed by	Description of Tasks Performed during Rescheduling
Dispatcher (using the STEG system [13])	*Computing solutions:* The dispatcher manually performs retiming, local rerouting and reordering of the trains by modifying the digital graph in STEG that depicts the current operational plan. *Selecting a solution:* The system shows the consequences of potential decisions by illustrating and comparing the rescheduled timetable with the original timetable. The dispatcher can accordingly reschedule and resolve the conflicts to obtain the preferred rescheduling solution.
Algorithm of Bettinelli et al. [14] (part of ICONIS system)	*Computing solutions:* Conflicts are resolved by reordering trains and retiming (i) the durations of train stops at stations, (ii) trains' entry and departure times in different parts of the network. Trains are also rerouted locally (e.g., platform changes) and globally, based on a predefined set of detours available for each train [14]. *Selecting a solution:* The rescheduled timetable with the least possible number of conflicts is presented to the dispatcher. The dispatcher considers the timetable output by the algorithm and accordingly selects a solution to be implemented.

3. Related Work

Train rescheduling algorithms and solution approaches have been reviewed time and again, e.g., [6,9,15]. In one of the early works, Törnquist [15] presents a review of algorithms and models for railway scheduling and dispatching. The author presents a framework to classify and compare in detail the various train scheduling approaches. More recently, Cacchiani et al. [9] present an overview of algorithms and recovery models for real-time railway disturbance and disruption management. Fang et al. [6] classify and compare the problem models, solution approaches, and problem types for rescheduling in railway networks. In these works, the important characteristics of various algorithmic approaches have been discussed and classified.

Practitioners and researchers may want to simultaneously compare outputs of two or more algorithms in order to assess their relative efficiency [16]. Limited research has been performed on comparing the performance of train rescheduling algorithms [6,17]. In one of the early works, Wegele et al. [7] compare two decision support tools, developed for the Dutch and German railway networks, to assess their effectiveness in optimal train rescheduling. The two train rescheduling algorithms use reordering as the rescheduling tactic and are configured to minimize the total train delays. Based on common input railway instances from the Dutch railway network, the authors propose a comparison between the obtained rescheduling solutions. The two algorithms and their obtained solutions are compared using (i) blocking time plots, (ii) total train delays and (iii) total travel time of all trains. The authors point out that their comparison is slightly imbalanced since the two tools model the train dynamics differently.

Min et al. [18] propose a train rescheduling algorithm, which they compare with the MILP-based heuristic algorithm of Törnquist and Persson [19] and a priority-based heuristic algorithm. The authors use real-world instances from the Seoul metro rail network comprising 23 stations and mixed railway traffic. The authors primarily compare and report (i) the objective values in the obtained solutions, (ii) the distribution of the relative optimality gap of the obtained solutions. The authors consider two cases: (a) the algorithms run to completion, (b) a predefined computational time limit of 1 min. However, the focus of their work is not on a comparison framework for train rescheduling algorithms. From a performance comparison point of view, a noticeable drawback in [18] is the lack of consideration of several other important quality indicators in the rescheduled timetables.

Fan et al. [17] compare eight different approaches (brute force, tabu search, simulated annealing, etc.) to solve the train rescheduling problem. The eight algorithms are configured to minimize the delay costs. The authors use (i) a rail infrastructure bounded by two simple junctions, (ii) a timetable consisting of 12 trains with mixed railway traffic and (iii) four disturbance scenarios, to evaluate the algorithms. The metrics used for evaluating the algorithms are (i) the ordering of trains, (ii) delay cost of each train, (iii) total delay cost (in pounds) and (iv) computation time. The authors comment on the suitability of the algorithm to solve a particular type of disturbance. It is unclear how the authors' approach can be extended to a larger infrastructure.

Samà et al. [16] evaluate several alternative MILP formulations of the train rescheduling problem with different objective functions. Their study focuses on (i) identifying the MILP formulations that give inefficient solutions and modifying them with the addition of appropriate constraints and (ii) identifying relatively efficient formulations among a set of available formulations. They perform experiments on a Dutch railway network with mixed traffic and multiple delayed trains, using rescheduling time windows of 30 min and 1 h.

In recent times, researchers have surveyed and discussed the different objectives and quality indicators for railway rescheduling in various contexts, e.g., Samà et al. [16], Törnquist Krasemann [20], Corman et al. [21], Josyula et al. [22]. While solving the train rescheduling problem, there is no general agreement in the literature on the objective function(s) to be adopted [16]. Similarly, often, there is no shared meaning for many quality indicators [21]. One example is the passenger inconvenience caused due to a rescheduled timetable, for which the literature adopts a wide range of definitions [23].

Based on the review of related work, some of the observed weaknesses and challenges are addressed in this paper by presenting: (i) a framework to evaluate and compare train rescheduling algorithms while using multiple quality indicators and a (ii) a proof-of-concept of the framework by comparing two multi-objective rescheduling algorithms. The two algorithms are extended versions of existing train rescheduling algorithms. The main contributions of the research presented in this paper are: (i) an evaluation framework for train rescheduling algorithms and a demonstration of its applicability and (ii) a systematic evaluation of the rescheduling solutions resulting from the two algorithms for realistic input data.

4. Framework Part I: Classification of Algorithm Capabilities and Characteristics

This section presents the first part of the evaluation framework, which serves to describe and compare alternative train rescheduling algorithms on a functional (or a conceptual) level. Originating from the existing classifications of train rescheduling algorithms in [6,9,15], we use the algorithm characteristics presented in Table 2 for the classification and description of algorithms for train rescheduling.

Many of the characteristics mentioned in Table 2 are elaborated in detail as follows.

Infrastructure granularity: A railway network can be considered on three different levels of granularity: microscopic, mesoscopic or macroscopic [11,14]. A microscopic modelling approach represents every relevant element of the railway infrastructure in detail, e.g., block sections of different length separated by signals and switches, properties of individual tracks and platforms in stations. This is typically important for scheduling the interaction of many different trains in congested sub-networks, stations and junctions. A macroscopic approach typically disregards any fine-grained segmentation of the tracks [14] and each modelled infrastructure element could represent several physical resources. For example, the capacity restriction of a segment between two stations is often represented by a cumulative function that restricts the number of trains that simultaneously occupy the segment, but without allocating unique tracks and platforms. Several algorithms adopt a mixed approach by using a mesoscopic model of the infrastructure and traffic [14]. Often, individual tracks and platforms are modelled, but not the layout of stations and junctions.

Time representation: Time representation refers to how the time that a train is scheduled to occupy a certain infrastructure resource is modelled. The choice of time representation affects how detailed the interaction of trains can be modelled and how the problem size grows with an increased scheduling time window. In Table 3, four rescheduling approaches that adopt a continuous time representation are mentioned, while e.g., Harrod and Schlechte [24] present and compare two alternative models that adopt a discrete time representation.

Special considerations: While rescheduling trains, a few core constraints need to be enforced for the feasibility of the resulting timetable(s). From a macroscopic modelling perspective, the following constitute the core constraints in train rescheduling:

1. Network capacity constraints: At most, one train can occupy a railway track at any time.
2. Minimum occupation time constraints: On a line section, a train may be able to run faster or slower than originally planned, but never run faster than the minimum run time for that specific section. On station sections, this corresponds to the minimum required dwell time.
3. Departure time constraints: A train, which stops at a station for alighting passengers, cannot depart that station before its originally planned departure time.

Table 2. The characteristics used to classify and describe algorithms and their capabilities.

Algorithm Characteristic	Values
Infrastructure granularity [9,11]	Microscopic, Mesoscopic, Macroscopic
Time representation [15]	Discrete, Continuous
Special considerations [15]	Train length, train weight, type of train service and its associated preferences, train connections and other operational dependencies, etc.
Applicable infrastructures [15]	Line, Network
Applicable sections [15]	Single-tracked, Double-tracked, Multi-tracked
Applicable railway tracks [15]	Unidirectional, Bi-directional
Rescheduling tactics [6,25]	Retiming, rerouting, reordering, train cancellations, adding extra trains, etc.
Optimization objective(s)	Minimize train delays, maximize passenger satisfaction, etc.
Solving strategy	Centralized, Decomposition
Solution space exploration	Full, Partial
Solution approach [6]	MILP solver, lagrangian relaxation method, branch and bound method using a depth-first search or a parallel tree search, tabu search, etc.
Main ideas of the approach [6]	Interpreting the problem as a blocking job-shop scheduling problem and modelling using graph theory, addressing multiple objectives in the problem using epsilon constraints, etc.
Control loop [26]	Open, Multiple open, Closed
Evaluation level [15]	Conceptual approach, Simulated experiments (artificial or real data), Field experiments, Deployed in practice
Evaluation context	Station or Terminal area, Line, Network
Applicable scenarios	Delayed train, infrastructure (e.g., signal) failure, train malfunction, freight train's early departure from its yard, etc.

Naturally, depending on if the problem is modelled using a microscopic, mesoscopic, or macroscopic approach, the core constraints are formulated in different ways. For example, in a mesoscopic approach, the headway and clear time constraints may be used to implicitly enforce the network capacity constraints. In such an approach, a track can be divided into several block sections and other units of physical track resources (e.g., switches).

If the algorithm accounts for other problem characteristics and constraints besides the core constraints, they are mentioned under the special considerations [15]. Examples of such constraints are (i) consideration of train length when allocating a platform for passengers transfer, (ii) considering synchronized arrival and departures of connecting trains, and (iii) considering the length and/or weight of freight trains when rescheduling unplanned stops and overtakes, which may introduce additional constraints [27].

Applicable infrastructures, sections and tracks: These three related aspects specify the properties that can be represented by an algorithm's model of the rail infrastructure. For example, an algorithm that is intended to be used for rescheduling trains on a single line may assume that all stations are linked in a sequence and that all trains travel between the stations in a chronological order. A line is here a "sequence of segments between two major stations with possibly several intermediate stations", while a railway network is instead comprised of "one or several junctions of lines" [15]. In a network, one station can be connected directly with more than two stations [6]. Hence, an algorithm may or may not be able to reschedule trains in a network setting. Furthermore, the segments between the stations and their capacity restrictions may be modelled differently depending on if e.g., only single track is considered or if segments with several alternative tracks can be used to reschedule delayed trains. Whether the algorithm can represent only tracks that permit traffic in only one direction, or in both directions (i.e., bi-directional traffic), is also relevant to capture. For example, in some rail networks, a double-tracked line consists of two parallel, uni-directional tracks where one is dedicated to traffic in one direction and the other to traffic in the reversed direction. Hence, overtaking is then

only possible at stations and an algorithm may base its computation on this assumption and enforce associated constraints. In Sweden, basically, all tracks allow bi-directional traffic (i.e., there are signals for trains in either direction). Allowing faster trains to overtake slower trains on the line between stations is a frequently-used measure to enable trains to catch up and to reduce delay propagation.
Solving strategy: When the original problem is solved as one instance, it is said to be a centralized approach [15]. A *decomposition* approach replaces the original problem with a sequence of smaller sub-problems, the solutions to which are computed and then recombined or extended to the original problem. Examples are (i) the rolling horizon approach (decomposition in time), (ii) partitioning trains into groups and sequentially solving the problem associated with each group, and (iii) approaches where entire administrative areas are considered as single entities to carry out inter-area coordination among trains [25].
Solution space exploration: Based on the country, IMs have specific rules for resolving conflicts during a disturbance. For example, in Sweden, the general dispatching strategy gives priority to on-time trains over the trains that deviate from the originally planned schedule. The reason behind this rule is to prevent a delay from propagating to trains that run according to schedule [28]. The dispatchers can, however, make exceptions to this rule when it is well-motivated. An algorithm that abides by such specific rules cannot fully explore the solution space for all possible rescheduling solutions.
Control loop: A *control loop* gives the interaction between the rescheduling tool and traffic operations [26]. In *open loop* rescheduling, the rescheduling decisions are computed and implemented only once at the beginning of a selected time window (e.g., two hours from the time when disturbance occurs). In *multiple open loop* rescheduling, the algorithm can be applied at successive times over the time window. Whenever additional information regarding traffic conditions is available, the calculations can be reconsidered [11]. However, the algorithm does not consider the actions computed and implemented during its previous runs in the selected time window. A closed loop rescheduling is defined as a multiple open loop with memory [26]. In this type of control loop, dispatching actions are immediately computed and adjusted every time updated information is available, on the basis of the current traffic state and the previously computed rescheduling decisions. In a closed loop, information updates are taken into account whenever available [11].
Evaluation context: The context in which the algorithm designer evaluates his/her proposed algorithm can be classified as: a station or a terminal area, a line, or a network.
Applicable scenarios: An algorithm could be designed such that it is applicable only to a subset of the possible disturbance scenarios. In contrast, an algorithm may be able to solve any type of disturbance scenario. An algorithm with the latter functionality could be more relevant in a practical context, where any type of disturbance scenario could arise. The applicable scenarios include the types of disturbances that an algorithm has been demonstrated to be able to solve.

Table 3. Algorithm characteristics of four train rescheduling algorithms.

Algorithm Characteristic	Algorithm 1 (Josyula et al. [22])	Algorithm 2 (Törnquist and Persson [19])	Algorithm 3 (Bettinelli et al. [14])	Algorithm 4 (Lamorgese and Mannino [29])
Infrastructure granularity	Mesoscopic	Mesoscopic	Configurable to any of the three granularity levels.	Mesoscopic: stations (microscopic), lines (macroscopic)
Time representation	Continuous	Continuous	Continuous	Continuous
Special considerations	Platform and track allocation of trains without considering train and track properties.	Platform and track allocation based on train length and track length. Train connections.	Connecting trains, (de)coupling of trains, track length, maximum train speed attainable on the track.	Train properties are considered in some cases.
Applicable infrastructures	Network	Network	Network	Network
Applicable sections	multi-tracked	multi-tracked	multi-tracked	multi-tracked
Applicable railway tracks	bi-directional	bi-directional	uni-directional	uni-directional, can be made applicable to bi-directional tracks by adding multi-commodity flow variables and constraints.
Rescheduling tactics	Retiming, reordering, and local rerouting	Retiming, reordering, and local rerouting	Retiming, reordering, and rerouting (local, global)	Retiming, reordering, and local rerouting
Optimization objective(s)	Minimize total train delay, total passenger delay, number of delayed passengers, etc. Objectives are easily configurable. Each objective corresponds to an objective function.	(i) Minimize train delays at their final destination or (ii) Minimize the total cost associated with train final delays. The two alternative objective functions can easily be extended and replaced.	Minimize the penalties due to (i) delays, (ii) logical dependency breaking, (iii) soft capacity violations, and (iv) the use of detours. The objective function can be easily modified to consider other indicators.	Minimize the sum of delay costs for all trains at every station in their route. The objective function can be made to accommodate various goals by adding specific variables.
Solving strategy	Centralized	Centralized	Centralized	Decomposition

Table 3. *Cont.*

Algorithm Characteristic	Algorithm 1 (Josyula et al. [22])	Algorithm 2 (Törnquist and Persson [19])	Algorithm 3 (Bettinelli et al. [14])	Algorithm 4 (Lamorgese and Mannino [29])
Solution space exploration	Partial	Full	Partial	Full
Solution approach	Branch and bound approach using a parallel depth-first search.	Branch and cut approach using a sequential tree search (defined by the applied commercial solver).	An iterated greedy approach using a reduced variable neighbourhood search, or tabu search.	Branch and cut approach (CPLEX solver to solve the formulated MILPs) using a parallel tree search.
Main ideas of the approach	A set of upper bounds associated with the set of best solutions is maintained and updated throughout the search. Pruning metrics corresponding to the multiple minimization objectives are used to prune off potentially undesirable solution branches.	The explicit MILP formulation is solved using a commercial solver, e.g., CPLEX or Gurobi. Optionally, the solution space can be restricted with different heuristic strategies to speed up the computation time while risking to cut solution branches leading to optimal solutions.	Each train's new schedule is created by solving a shortest-path problem. A new timetable is obtained by rescheduling the trains one by one, according to their ranking, which is perturbed using a set of rules to obtain improved timetables.	A master–worker algorithm is applied to the line (as master) and station (as worker) subproblems, by modelling them as MILPs.
Control loop	Open	Open	Open	Open
Evaluation level [15]	Simulated experiments (real-world instances of Sweden).	Simulated experiments (real-world instances).	Simulated experiments (real-world and artificial instances). (To be) deployed in practice.	Simulated experiments (real-world instances of Italy). Deployed in practice in Norway.
Evaluation context	Line (Karlskrona–Tjörnarp line)	Network (of different lines in the south traffic district of Sweden).	Line, Station area	Line (Lines Stavanger-Moi, Foligno-Orte, etc.)
Applicable scenarios	Delayed train (i.e., initial primary delays of various sizes, various trains and various locations).	Delayed train (i.e., initial primary delays of various sizes, various trains and various locations).	Cannot be determined from the research paper.	Cannot be determined from the research paper.

We apply the aforementioned part of the evaluation framework on four train rescheduling algorithms to compare their capabilities. The classification of the four algorithms, shown in Table 3, is based on the descriptions and demonstrations of the approaches in the mentioned references. Hence, detailed information is not available to cover all aspects to the same extent. Furthermore, other versions of those approaches may also exist and be in use in other settings.

5. Framework Part II: Key Aspects for a Systematic Evaluation of Algorithm Performance

This section presents the second part of the framework, which suggests a selection of key aspects to be used in systematic performance evaluation and benchmarking of algorithms.

During train rescheduling, the objective(s) of an algorithm refer to the aspects that are to be minimized and/or maximized in the solutions [5]. A train rescheduling algorithm may produce one or more rescheduling solutions. The objectives that drive the computation of solutions indicate the quality aspects that are important to be considered by the algorithm. However, there may also be other properties of the produced solution(s) that affect their relevance and acceptability, of which some properties may be difficult to incorporate explicitly in the computations of good rescheduling solutions. For example, an IM may want to assess the robustness of the timetables produced by the algorithm. This property is typically easier to define and compute once the solutions are generated, but less suitable to formalize as a constraint or penalty.

Having a standard set of quality indicators allows comparison of solutions computed by different algorithms, irrespective of the objectives of the specific algorithms. A quality indicator may be comprised of one metric or a set of metrics. These metrics can be used to compare algorithms and to reveal their strengths and weaknesses. These can also be used to monitor the search process of an algorithm and explicitly guide the search for improved solutions [30]. Table 4 lists the seven quality indicators and their corresponding metrics that are considered in the proposed evaluation framework.

Table 4. A description of the chosen quality indicators.

Indicator	Employed Metrics
Train punctuality	(i) The percentage of early and on-time trains, (ii) the percentage of delayed trains for various thresholds.
Train delays	The total final and accumulated delays in minutes with a threshold value of three minutes, i.e., TFD_3, TAD_3, and their closeness to the hypothetical, ideal point.
Delay propagation	The number of trains with secondary delays, considering a threshold value of three minutes. That is, the number of trains with secondary delays ≤ 3 min and >3 min.
Freight train performance	(i) Deviations in departure times of freight trains at their yards (in min), (ii) increase in freight train travel times (in min), (iii) number of unplanned stops for freight trains, (iv) the percentage of freight trains that arrive earlier than a threshold value (of 15 min) at their arrival yards.
Passenger delays	Total passenger delay (in minutes) exceeding a threshold of three minutes, i.e., TPD_3, and its closeness to the optimal value.
Track reassignments	The number of rescheduled track allocations for passenger and freight trains, at stations and line sections.
Computation time	The wall-clock time taken by the algorithm to obtain the rescheduled timetable.

1. *Train punctuality:* The percentage of trains that arrive at their final destination within a given threshold of t minutes represents train punctuality. This metric is frequently used by railway organizations and in rail literature, with various threshold values, e.g., 0 min [16], 3 min [20,29], 5

min [21]. For this reason, we use the percentage of early, on-time and delayed trains (for different threshold values) as metrics for train punctuality.

2. *Train delays:* Tardiness of a train at a relevant point in the network, e.g., a station, is its delay in arriving at the point [16], within a chosen t min threshold. Total accumulated delay (TAD) is the total tardiness of all trains at their intermediary, scheduled commercial stations (a commercial station is a station where the train stops for alighting passengers). Total final delay (TFD) is the total tardiness of all trains at their final destinations.
   The tardiness metrics TAD_3 and TFD_3 capture the delays for a threshold of 3 min. The delays incurred in the solutions output by an algorithm are expected to be as close as possible to the optimal.
   Total accumulated delay is an important delay metric often used in railway operations analysis [31]. Total final delay is a frequently used metric in the objective functions of existing train rescheduling algorithms proposed by the research community. The reason for selecting a threshold of 3 min is due to its use by the Swedish railway authority to continuously monitor and log arrival and departure train delays in the associated traffic management system. Furthermore, delays larger than 3 min are more likely to cause the passengers to miss train connections, compared to smaller delays.
3. *Delay propagation:* Trains that are delayed as a direct result of a disturbance experience primary delays. When on-time trains instead become delayed as a result of delay propagation, they are said to experience secondary delays. We compute the number of trains experiencing small secondary delays ($\leq$3 min) and large secondary delays ($>$3 min) anywhere in their itinerary, in the obtained rescheduled timetables. This indicator is used to reflect the extent to which delays are propagated to other trains (i.e., knock-on effects). It is important to consider this indicator as even a simple infrastructure failure can at times create knock-on effects in the railway network that may continue for many hours.
4. *Freight train performance:* The timetable deviation [16] evaluates the difference between the originally planned timetable and the new timetable. The latter should ideally limit the deviation from the originally planned times. We compute the timetable deviation for freight trains at their departure yards. Since freight train operators prefer not to increase their travel times, we consider it to be a relevant metric to use when measuring freight train performance. Another important metric for freight trains is the number of unplanned stops [20]. Multiple unplanned stops further increase the travel time, since slowing down and speeding up heavy freight trains is time consuming. In addition, these trains may block critical station tracks during their unplanned stops. In addition, unplanned stops impact operating costs and energy consumption. Hence, we record the number of new stations at which freight trains stopped during their journey. When a freight train arrives very early, problems may arise in the arrival yard where shunting and (un)loading takes place as per its separate schedule. For this reason, we use the percentage of freight trains that are earlier than 15 min as a metric for this indicator. Late arrivals of freight trains are also problematic. The percentage of late freight trains, e.g., with an arrival delay $\geq$ 30 min, are already accounted for in the general train punctuality indicator, and hence we do not explicitly record them.
   Note that, if the rescheduling time window does not include the freight train's complete journey, the first and last stations in the problem instance are considered to be the departure and arrival yards, respectively.
5. *Passenger delays:* Total passenger delay (TPD) captures the total delays experienced by all passengers while alighting at their destinations. This metric is frequently used in the rail literature to estimate passenger inconvenience [23]. We multiply the number of alighting passengers with the associated train delay at that particular train stop, where only a delay larger than a threshold of 3 min is counted. This metric, called TPD_3, is used to estimate the inconvenience that the passengers would experience due to the rescheduled timetable.

6. *Track reassignments:* The number of track reassignments indicates how complicated a rescheduling solution may be to implement. Furthermore, the track reassignments for a passenger train at a station may result in certain passenger groups having to change platforms shortly before the train's departure. In practice, this may result in either the passengers missing their train or in a new train delay due to increased boarding times. Note that passenger transfers and train coordinations are not considered in the two algorithms compared in this paper.
7. *Computation time*: The computation time of a train rescheduling algorithm is an important metric, as the algorithm runs in real time. The wall-clock time taken by the algorithm to obtain the best found rescheduling solution is recorded. Alternatively, an appropriate time limit (e.g., 15 seconds) could be set for the benchmarked algorithms to assess the resulting best rescheduling solutions within the time limit.

6. Train Rescheduling Algorithms Used in the Experiment

We conduct an experiment on two alternative algorithms and apply the proposed framework to assess and compare the performance of each algorithm. In this section, we describe the two chosen algorithms.

When considering the perspectives of multiple stakeholders, solving the train rescheduling problem with separate multiple objectives may be more beneficial and natural than other approaches, e.g., the weighted sum approach [6]. An a priori method for multi-objective optimization requires the preference information regarding the objectives to be expressed prior to the solution process [32], e.g., the lexicographic method [33]. In contrast, an a posteriori method returns a solution set which is a representative of the Pareto-optimal solutions. The final solution is then chosen from the available set, either by using another method or by the decision maker. Below, we present a short description of the two train rescheduling algorithms that are evaluated and benchmarked in the experiment.

6.1. Algorithm 1: An a Posteriori Multi-Objective Parallel Heuristic Algorithm

Algorithm 1 (denoted ALG1 hereafter) is an extension of the multi-objective parallel heuristic algorithm presented in [22]. The algorithm constructs (and simultaneously navigates) a binary tree by iteratively detecting and resolving conflicts. The root node corresponds to the original timetable which turns infeasible due to the disturbance. At each node, a conflict detection operation is performed on the corresponding timetable. The detected conflicts are arranged in a chronological order and the first conflict is chosen to be resolved. Each internal node of the binary tree represents a conflict between exactly two trains. Each outgoing edge corresponds to the rescheduling decision made as a part of conflict resolution. Reordering, retiming trains, and local rerouting are the employed rescheduling tactics. Leaf nodes in the unpruned branches correspond to feasible solutions.

The search tree construction is decomposed into disjoint tasks, which are computed in parallel by multiple threads. Starting with the root node, each node is visited using a parallel depth-first search (DFS) strategy to find the best solution. A selection of evaluation metrics (e.g., TFD_3, TPD_3) are used for pruning. Throughout the search, all the parallel threads share and update the values of the upper bound of each selected metric. Based on these values, the branches leading to undesirable solutions are pruned. The parallel program can be launched with the required number of parallel threads. Once the specified number of threads are created, each thread runs in parallel an instance of the sequential DFS.

In any intermediate timetable state, i.e., at any internal node, first, one of the trains in the chosen conflict is prioritized over the other. Typically, each of the two outgoing edges corresponds to a prioritization alternative. Then, a child node is created by performing the following actions: (i) by locally rerouting the unprioritized train if an empty track is available throughout the train's occupancy of the conflict section, (ii) otherwise, by making the unprioritized train wait on a prior section (likely causing a reordering), and retiming it accordingly to resolve the conflict. Thus, reordering is always accompanied by retiming. Each edge in the binary tree, i.e., each rescheduling decision, corresponds to either (i) a track reassignment, (ii) retiming, or (iii) reordering and retiming, of a train. See [22] for further details about the algorithm.

The following extensions to [22] have been done to construct ALG1:

1. Limiting the number of track reassignments: While rescheduling, the algorithm in [22] often performed many track reassignments. Limiting of track reassignments is achieved by reallocating tracks only for the trains with a primary delay. For all other trains, no track reassignments are performed; reordering and retiming are the employed tactics.
2. Improving the resolution of conflicts: The algorithm in [22] (i) does not make use of the buffer times available in the initially disturbed events, (ii) uses only the FCFS prioritization strategy to resolve a conflict that involves two trains in the same direction, (iii) does not have any memory of previously resolved conflicts. The conflict resolution module of the algorithm in [22] is redesigned (i) to use the buffer times in the train events disturbed due to the incident, (ii) to consider both the prioritization alternatives when a conflict between two trains in the same direction is encountered and (iii) to consider the same prioritization between two trains in conflict throughout a solution branch.
3. Implementing the TOPSIS approach: The multi-objective algorithm in [22] returns a set of rescheduled timetables. Several useful methods exist to select one solution from a set of rescheduling solutions. One such method is TOPSIS [34], which stands for Technique for Order of Preference by Similarity to Ideal Solution. This method is used to select the best train timetable from the set of timetables output by the multi-objective algorithm. The TOPSIS method selects an alternative such that it is closest to the ideal solution and farthest from the negative-ideal solution.
4. Speeding up the search by ignoring potentially undesirable solutions: During the search for a set of good candidate solutions, the multi-objective algorithm in [22] searches along a solution branch whenever the partial solution improves an existing upper bound, even by a minute amount. We noticed that this increases the computation time of the algorithm and gives undesirable solutions in the solution set which are later excluded by the TOPSIS approach. Hence, the algorithm's search process is modified such that it explores a solution branch only when the partial solution improves an existing upper bound by at least 20% in any of the metrics considered in the objectives.

Table 5 summarizes the main differences between ALG1 and the multi-objective algorithm in [22]. In the experiment, ALG1 is configured to consider two objectives: minimizing TFD_3 and TPD_3. It is run in parallel using eight threads, and equal weights are used in the TOPSIS method.

Table 5. Main differences between the multi-objective algorithm in [22] and ALG1.

The Multi-Objective Algorithm in [22]	The Extended Algorithm (ALG1)
Reallocates tracks whenever possible.	Reallocates tracks only for the trains with a primary delay.
Does not modify the train events disturbed due to the incident.	Uses the buffer times in the train events disturbed due to the incident.
Applies only the FCFS rule while prioritizing and retiming the trains, whenever a node has a conflict between two trains in the same direction.	Considers both the prioritization alternatives when a conflict between two trains in the same direction is encountered.
Considers any prioritization between two trains in conflict throughout a solution branch.	Considers the same prioritization between two trains in conflict throughout a solution branch.
Collects potentially undesirable solutions in the solution set.	Ignores potentially undesirable solutions during the search process.

6.2. Algorithm 2: An a Priori Multi-Objective Optimization Model Solved Using a Commercial Optimization Software

Algorithm 2 (denoted ALG2 hereafter) is a lexicographic extension of the single-objective event-based MILP model described in [35] and originally proposed by Törnquist and Persson [19]. The MILP model in [19] has more restrictions than the slimmed-down version (i.e., ALG2) used for our benchmarks in the experiment. The two algorithms, ALG1 and ALG2, use the same constraints and problem formulation.

When using the lexicographic method, preferences of the objectives are imposed by ordering the objective functions according to the practitioner's choice, rather than by assigning weights. The advantages of this method are that it does not require the objective functions to be normalized, and it always provides a Pareto-optimal solution as output [33].

The Java code corresponding to the implementation of the MILP model is extended by adding the code in Listing 1, thus making it a multi-objective algorithm. The `setObjectiveN()` method is used to set five objectives with different priorities. In Listing 1, a unique integer `index` in $[0, n-1]$ is assigned to each of the n objectives, as required by the employed commercial solver. The integer `priority` for each objective is assigned, keeping in mind that the larger the value, the higher is the priority. The solver allows lower-priority objectives to degrade those with a higher priority by the specified absolute or relative tolerance amounts (`abstol` or `reltol`, respectively). In our lexicographic approach, we restrict lower-priority objectives from degrading the values of higher-priority objectives by specifying the values of `abstol` and `reltol` as zero.

Listing 1. Code that corresponds to adding multiple objectives.

```
// setObjectiveN(expression, index, priority, weight, abstol, reltol, name)
// Primary objective and four other objectives
model.setObjectiveN(tfd3, 0, 4, 1, 0, 0, "TFD+3min");
model.setObjectiveN(tad3, 1, 3, 1, 0, 0, "TAD+3min");
model.setObjectiveN(tpd3, 2, 2, 1, 0, 0, "TPD+3min");
model.setObjectiveN(track_reassignments, 3, 1, 1, 0, 0, "Track reassignments");
model.setObjectiveN(event_end_deviations, 4, 0, 1, 0, 0, "Deviations");
```

After extending the implementation of the MILP model by using `setObjectiveN()` with the appropriate arguments, we employ the commercial solver, which uses the following algorithm (called ALG2 in this paper), to solve the extended model. ALG2 first searches for an optimal solution for the highest-priority objective, i.e., minimizing TFD_3. It then searches for an optimal solution for the next objective [36], i.e., minimizing TAD_3, but only from among the solutions with optimal value of TFD_3. The algorithm then searches for an optimal solution that minimizes TPD_3 from among the solutions with optimal TFD_3 and TAD_3. Similarly, the algorithm searches for an optimal solution that minimizes track reassignments. Finally, the algorithm finds an optimal solution that minimizes event end-time deviations and outputs that solution to the user. Note that the relative prioritization between the five objectives in ALG2 is not connected in any way to the proposed framework.

7. Description of the Experiment

In Section 7.1, we describe the input dataset and the disturbance scenarios used in the experiment. In Sections 7.2–7.4, we describe the experimental setup, the gathering of ideal points, and the selected statistical test, respectively. Finally, in Section 7.5, we discuss the fairness in benchmarking the algorithms.

7.1. Dataset and Scenarios

A railway network in the southern part of Sweden is chosen for the experiment. The chosen network comprises the railway stretch between Karlskrona-Malmö, via Kristianstad and Hässleholm (see Figure 1). The railway line is (i) single-track from Karlskrona to Hässleholm, (ii) double-track from Hässleholm to Malmö, (iii) with four tracks between Arlöv and Malmö. The original timetable consists of mixed traffic. It includes (i) regional passenger trains that take a travel time of 1.5 h between Karlskrona and Kristianstad, and 1 h between Kristianstad and Malmö, (ii) freight trains that run different stretches, (iii) long-distance passenger trains with a piece of their journey between Hässleholm-Malmö. Table 6 presents the characteristics of the problem dataset used in the experiment.

The 30 disturbance scenarios constituting the dataset are described in Table 7. All of them occur during peak hours: 4:00 p.m.–6:00 p.m. A rescheduling time window of 1.5 h is considered. The time window starts from the time of occurrence of the disturbance.

In the first ten disturbances, a passenger train initially experiences a primary temporary delay (of 7–25 min) at one section within the infrastructure. In each of the next ten disturbances, a passenger train has a malfunction, resulting in increased minimum running times (between 20–100%) on all sections it plans to occupy. For these scenarios, the percentage increase in the minimum train running time, e.g., 20%, is mentioned. In the final ten scenarios, the disturbance is due to an infrastructure failure causing, e.g., a speed reduction on a particular section, which results in increased minimum train running times (of 2–6 min) for all trains running through that section. Table 7 shows, for each disturbance, the total number of trains with a primary delay. For the last ten disturbance scenarios, the number of freight trains incurring initial primary delay is also mentioned.

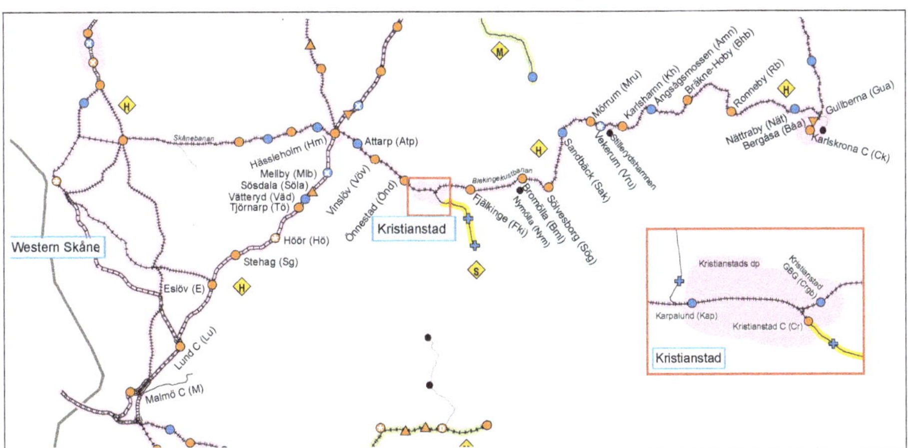

Figure 1. Considered infrastructure: Karlskrona-Malmö [37].

Table 6. Characteristics of the problem dataset.

Characteristic	Description
Type of infrastructure	Network
Number of sections	90, of which 42 are stations.
Number of block sections	290
Total number of trains	237, of which 198 are passenger trains and the remaining 39 are freight trains.
Passenger data	Synthetic data on the flow of passengers: At commercial stations, a random number of passengers (up to 85) board the passenger trains. Likewise, a random number of on-board passengers alight the trains. All of the remaining passengers on board a train alight at its final station.
Total number of passengers	11,545
Types of disturbances	Train delay at a section, train malfunction, infrastructure failure

Table 7. The dataset containing 30 disturbances.

Scenario	Disturbance Location (Tracks)	Initially Disturbed Train	Initial Delay	1.5 h Time Window		
				Total Trains (of Which Freight)	Initially Disturbed Trains (of Which Freight)	Events
1	Karlshamn-Ångsågsmossen (1)	1058 (Eastbound)	10 min	96 (17)	1	1619
2	Bromölla-Sölvesborg (1)	1064 (Eastbound)	15 min	88 (13)	1	1538
3	Kristianstad-Karpalund (1)	1263 (Southbound)	18 min	81 (11)	1	1341
4	Bergåsa-Gullberna (1)	1097 (Westbound)	20 min	89 (13)	1	1569
5	Bräkne Hoby-Ronneby (1)	1103 (Westbound)	25 min	85 (14)	1	1312
6	Flackarp-Hjärup (2)	491 (Southbound)	20 min	82 (9)	1	1359
7	Eslöv-Dammstorp (2)	533 (Southbound)	20 min	95 (17)	1	1631
8	Burlöv-Åkarp (2)	544 (Northbound)	7 min	95 (17)	1	1616
9	Burlöv-Åkarp (2)	1378 (Northbound)	20 min	81 (11)	1	1341
10	Höör-Stehag (2)	1381 (Southbound)	20 min	87 (15)	1	1595
11	Karlshamn-Ångsågsmossen (1)	1058 (Eastbound)	40%	96 (17)	1	1619
12	Bromölla-Sölvesborg (1)	1064 (Eastbound)	20%	88 (13)	1	1538
13	Kristianstad-Karpalund (1)	1263 (Southbound)	20%	81 (11)	1	1341
14	Bergåsa-Gullberna (1)	1097 (Westbound)	40%	89 (13)	1	1569
15	Bräkne Hoby-Ronneby (1)	1103 (Westbound)	100%	85 (14)	1	1312
16	Flackarp-Hjärup (2)	491 (Southbound)	100%	82 (9)	1	1359
17	Eslöv-Dammstorp (2)	533 (Southbound)	50%	95 (17)	1	1631
18	Burlöv-Åkarp (2)	544 (Northbound)	80%	95 (17)	1	1616
19	Burlöv-Åkarp (2)	1378 (Northbound)	40%	81 (11)	1	1341
20	Höör-Stehag (2)	1381 (Southbound)	40%	87 (15)	1	1595
21	Karlshamn-Ångsågsmossen (1)	All trains passing through	4 min	96 (17)	4 (0)	1619
22	Bromölla Sölvesborg (1)	All trains passing through	2 min	88 (13)	7 (0)	1538
23	Kristianstad-Karpalund (1)	All trains passing through	3 min	81 (11)	10 (0)	1341
24	Bergåsa-Gullberna (1)	All trains passing through	6 min	89 (13)	7 (0)	1569
25	Bräkne Hoby-Ronneby (1)	All trains passing through	5 min	85 (14)	4 (0)	1312
26	Flackarp-Hjärup (2)	All trains passing through	3 min	82 (9)	38 (5)	1359
27	Eslöv-Dammstorp (2)	All trains passing through	4 min	95 (17)	26 (5)	1631
28	Burlöv-Åkarp (2)	All trains passing through	2 min	95 (17)	46 (6)	1616
29	Burlöv-Åkarp (2)	All trains passing through	2 min	81 (11)	38 (4)	1341
30	Höör-Stehag (2)	All trains passing through	2 min	87 (15)	25 (5)	1595

7.2. Experimental Setup

The experiment is performed on a laptop equipped with a quad-core CPU (Intel Core i7-8550U(Santa Clara, CA, USA)) and 16 GB RAM. The underlying operating system is Windows 10 Education. (Redmond, WA, USA). The compiler used to compile the C++ code corresponding to ALG1 is Microsoft C/C++ Optimizing Compiler Version 19.14 for x64. The Gurobi Optimizer version 8.0.0 was used to solve the MILP model, with the default number of parallel threads (i.e., eight threads). ALG1 is also configured to run using the same number of threads.

7.3. Gathering the Ideal Point for Each Scenario

In order to assess the train and passenger delays in the best solutions computed by ALG1 and ALG2, we need to have a common reference. The performance of a solution in terms of optimality can be quantified by computing its closeness to the ideal point. We compute the ideal point for each disturbance scenario by using the optimization solver to optimize each objective individually. For example, in this case, by solving the rescheduling problem three times using the single-objective MILP model with the following objectives: (i) minimizing TFD_3, (ii) minimizing TAD_3, (iii) minimizing TPD_3. Typically, the ideal point is hypothetical, i.e., it often does not exist in the solution space [33]. Table 8 shows the computation of ideal point for disturbance scenario 1.

Table 8. Computing ideal point for disturbance scenario 1.

Description	TFD_3	TAD_3	TPD_3
Optimal solution with minimum TFD_3	1.1 min	-	-
Optimal solution with minimum TAD_3	-	8.1 min	-
Optimal solution with minimum TPD_3	-	-	305.7 min
Ideal point	1.1 min	8.1 min	305.7 min

7.4. Statistical Analysis

Given the results obtained for the input problem dataset, for a performance indicator, we want to confirm or reject statistically that there is a significant difference in the performance of the algorithms. To this end, the results of the experiment are analyzed using a statistical test. A general assumption of many statistical tests, called parametric tests, is that the data are normally distributed [38]. Non-parametric statistical tests, unlike their parametric counterparts, make no assumption about the data distribution. Hence, we use a non-parametric test called Wilcoxon signed-rank to test our hypotheses.

Using the Wilcoxon signed-rank test, we test if there is a difference between the performance of the two algorithms. The null hypothesis is that the median of the differences between pairs of observations is zero [39]. The p-value is interpreted as the probability that the difference in the medians of the observations (corresponding to the two algorithms) can be attributed to chance alone [40]. We apply the two-tailed Wilcoxon signed-rank test with a significance level $\alpha = 0.05$. We reject the null hypothesis if the obtained p-value is less than 0.05. Rejecting the null hypothesis shows that the difference between the performance of the two algorithms was unlikely to occur by chance. We use the `scipy.stats.wilcoxon()` function from the open-source software SciPy to perform the Wilcoxon signed-rank test. In the next section, wherever relevant, we mention the p-value obtained from the aforementioned test.

7.5. Fairness in Benchmarking the Algorithms

Beiranvand et al. [41] provide key recommendations for a fair benchmarking of optimization algorithms. Based on their recommendations, we described the algorithms, their parameters, the problem dataset, the computational environment, and the employed statistical techniques with an acceptable level of detail.

If the goal of algorithm comparison is to determine the best algorithm to use for a particular real-world application, using a real-world dataset is typically the best option [41]. The problem dataset used in our experiment is based on real-world data, in contrast to an artificial dataset. When benchmarking optimization algorithms, measuring wall-clock time is very relevant in real-world settings. The other alternative is to measure CPU time, which has its pros and cons [41]. In order to maximize the reliability of the collected data, we ensure that the background operations of the computer are kept to a minimum.

Many studies that compare optimization algorithms use basic statistics (e.g., average execution time) to report the experimental results [41]. Though it is reasonable to report those, a disadvantage is that such statistics provide little information about the overall performance of the compared algorithms. Numerical tables allow comprehensive reporting of benchmarking results and are recommended to be reported for the sake of completeness [41]. We report detailed numerical tables and analyze the results in a transparent and fair manner.

8. Results and Discussion

The application of the first part of the evaluation framework was demonstrated in Table 3 of Section 4. The table comprised four algorithms, of which the first two algorithms are the bases of ALG1 and ALG2, respectively. For most of the characteristics, both ALG1 and ALG2 have the same values as their base version algorithms (shown in Table 3). The remaining five characteristics are shown in Table 9.

Table 9. Algorithm characteristics of the two algorithms (abridged).

Characteristic	ALG1	ALG2
Special considerations	Platform and track allocation of trains without considering train properties or track properties.	Platform and track allocation of trains. Train length, track length, and train connections are not considered.
Optimization objectives	Minimizing TFD_3, TPD_3 are the two considered objectives.	Minimizing TFD_3, TAD_3, TPD_3, track reassignments, event end deviations.
Main ideas of the approach	A set of upper bounds is maintained and pruning is performed based on multiple metrics.	Objectives are prioritized and a lexicographic approach is used to find the best solution.
Evaluation context	Network	Network
Applicable scenarios	Delayed train, infrastructure failure, train malfunction.	Delayed train, infrastructure failure, train malfunction.

In this section, we demonstrate the application of the second part of the evaluation framework proposed in Section 5. The results from the experimental benchmark of ALG1 and ALG2, based on the evaluation framework, are presented and analyzed. The detailed results are shown in Figures 2–5 and Tables 10–14.

For each scenario, the table cells corresponding to the algorithm with a comparatively large value are highlighted in grey. The average values of the recorded metrics are shown using Tables 15–19.

In the solutions obtained by the two algorithms, train punctuality is shown in Figure 2 and Table 10. Train delays are recorded in Table 12 and compared to the optimal values. Delay propagation is shown in Table 14. This table records the number of trains experiencing secondary delays anywhere in their itinerary, in the obtained rescheduled timetables. Freight train performance is shown in Figures 3 and 4. Track reassignments and passenger delays are shown in Tables 11 and 13, respectively. Computation times of the two algorithms are shown in Figure 5.

When the delay in the obtained solution: (i) is within 1% of the optimal value, the corresponding cell is not highlighted, (ii) is within 20% of the optimal value, the table cell is highlighted in light grey and (iii) is greater than 20% of the optimal value, the cell is highlighted in grey.

16%
12%
8%
4%
0%
1 2 3 4 5 6 7 8 9 10 11 12 13 14 15 16 17 18 19 20 21 22 23 24 25 26 27 28 29 30
Disturbance scenario number
ALG1
ALG2

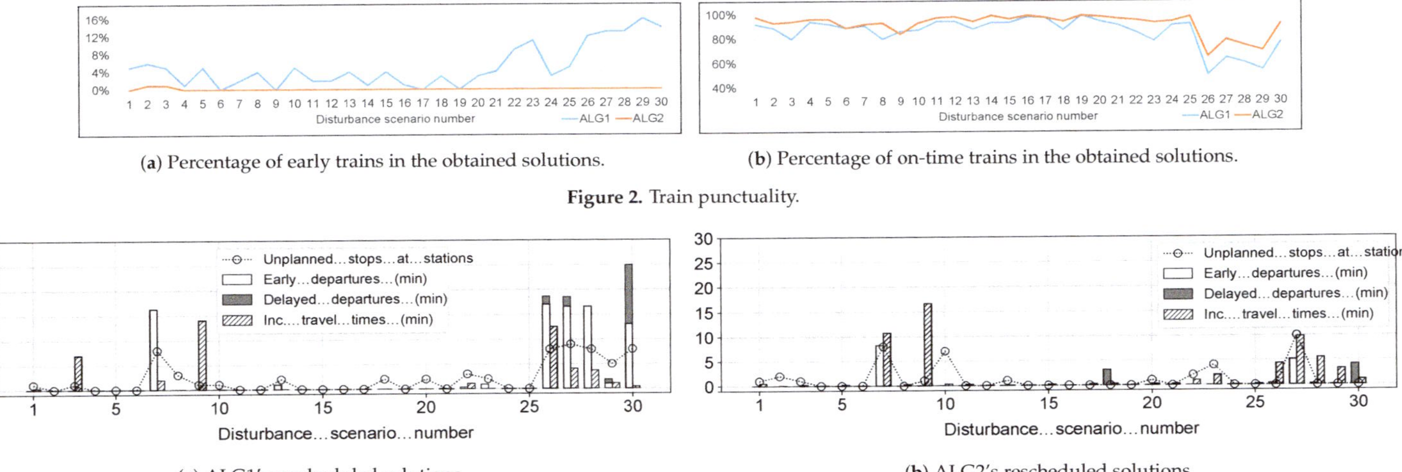

(**a**) Percentage of early trains in the obtained solutions.

100%
80%
60%
40%
1 2 3 4 5 6 7 8 9 10 11 12 13 14 15 16 17 18 19 20 21 22 23 24 25 26 27 28 29 30
Disturbance scenario number
ALG1
ALG2

(**b**) Percentage of on-time trains in the obtained solutions.

Figure 2. Train punctuality.

30
25
20
15
10
5
0
1
5
10
15
20
25
30
Unplanned...stops...at...stations
Early...departures...(min)
Delayed...departures...(min)
Inc....travel...times...(min)
Disturbance...scenario...number

(**a**) ALG1's rescheduled solutions.

30
25
20
15
10
5
0
1
5
10
15
20
25
30
Unplanned...stops...at...station
Early...departures...(min)
Delayed...departures...(min)
Inc....travel...times...(min)
Disturbance...scenario...number

(**b**) ALG2's rescheduled solutions

Figure 3. Freight train performance at departure yards and during the journey.

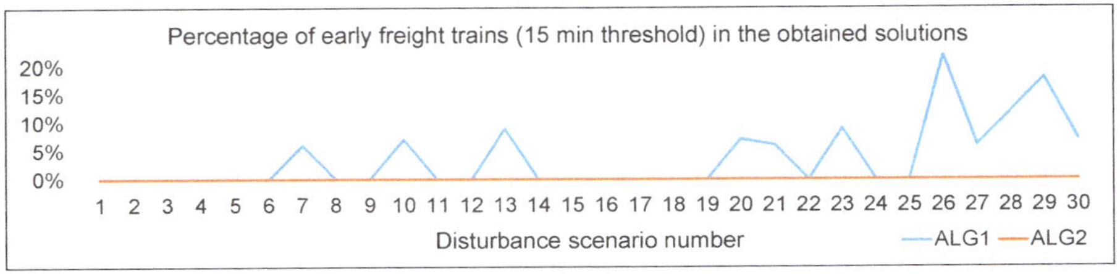

Percentage of early freight trains (15 min threshold) in the obtained solutions
20%
15%
10%
5%
0%
1 2 3 4 5 6 7 8 9 10 11 12 13 14 15 16 17 18 19 20 21 22 23 24 25 26 27 28 29 30
Disturbance scenario number
ALG1
ALG2

Figure 4. Freight train performance at arrival yards.

Table 10. Train punctuality.

Delays at Final Stations	Algorithm	Disturbance Scenario																													
		1	2	3	4	5	6	7	8	9	10	11	12	13	14	15	16	17	18	19	20	21	22	23	24	25	26	27	28	29	30
<3 min	ALG1	2%	3%	7%	-	2%	1%	2%	13%	4%	6%	3%	3%	6%	1%	1%	-	2%	8%	1%	1%	5%	6%	9%	-	1%	38%	20%	27%	30%	9%
	ALG2	1%	3%	2%	-	2%	4%	4%	5%	7%	5%	2%	2%	4%	-	1%	-	2%	2%	1%	1%	4%	5%	6%	-	1%	35%	18%	26%	30%	8%
[3, 10) min	ALG1	1%	1%	5%	3%	-	2%	3%	3%	4%	-	1%	-	2%	2%	-	1%	1%	-	-	1%	-	-	2%	6%	2%	-	3%	-	-	-
	ALG2	1%	2%	1%	3%	-	5%	3%	2%	4%	-	1%	-	2%	-	-	1%	1%	3%	-	1%	-	-	1%	6%	1%	-	3%	-	-	-
[10, 30) min	ALG1	-	1%	2%	1%	1%	7%	2%	-	6%	2%	-	-	-	2%	1%	-	-	-	-	-	-	-	-	-	-	-	-	-	-	-
	ALG2	-	-	1%	1%	1%	1%	1%	-	5%	2%	-	-	-	1%	1%	-	-	1%	-	-	-	-	-	-	-	-	-	-	-	-
≥30 min	ALG1	-	-	-	-	-	-	-	-	-	-	-	-	-	-	1% *	-	-	1%	-	-	-	-	-	-	-	-	-	-	-	-
	ALG2	-	-	-	-	-	1%	-	-	-	-	-	-	-	-	1% *	-	-	-	-	-	-	-	-	-	-	-	-	-	-	-

* Delays greater than 60 min.

Table 11. Track reassignments.

Algorithm	Disturbance Scenario																													
	1	2	3	4	5	6	7	8	9	10	11	12	13	14	15	16	17	18	19	20	21	22	23	24	25	26	27	28	29	30
ALG1	-	1	7	1	-	-	-	-	-	4	-	-	-	1	-	-	1	-	-	-	2	5	9	-	-	14	29	36	19	27
ALG2	-	3	3	1	-	-	-	-	2	1	-	-	-	1	-	-	-	3	-	-	-	-	3	-	-	-	-	-	-	-

Table 12. Train delays.

Scen	TFD_3 (min) in the Solution of			TAD_3 (min) in the Solution of		
	ALG1	ALG2	Optimal	ALG1	ALG2	Optimal
1	1.1	1.1	1.1	8.1	8.1	8.1
2	14.2	5.0	5.0	63.8	60.3	57.3
3	27.9	14.9	14.9	144.0	145.4	138.3
4	15.4	13.3	13.3	46.1	44.0	43.6
5	9.5	9.5	9.5	71.2	61.8	60.3
6	78.1	54.9	54.9	118.1	130.4	112.7
7	28.6	24.3	24.3	93.2	86.5	86.5
8	4.6	3.9	3.9	33.0	13.1	13.0
9	68.1	56.3	56.3	77.3	71.4	29.8
10	18.1	16.4	16.4	70.1	66.7	66.4
11	3.6	3.6	3.6	4.9	4.9	2.0
12	0.0	0.0	0.0	0.7	0.7	0.7
13	3.0	3.0	3.0	13.1	13.1	13.1
14	39.8	26.6	26.6	47.2	70.7	46.6
15	73.1	71.7	71.7	193.4	188.5	181.1
16	1.4	1.4	1.4	0.0	0.0	0.0
17	5.6	5.6	5.6	4.5	4.5	4.5
18	38.3	33.0	33.0	38.1	44.4	31.8
19	0.0	0.0	0.0	0.0	0.0	0.0
20	0.1	0.1	0.1	0.0	0.0	0.0
21	0.0	0.0	0.0	3.4	3.4	3.4
22	0.0	0.0	0.0	0.0	0.0	0.0
23	1.7	1.3	1.3	2.1	0.0	0.0
24	16.6	16.6	16.6	12.8	12.8	10.3
25	5.4	4.7	4.7	2.7	2.7	2.4
26	0.0	0.0	0.0	0.0	0.0	0.0
27	0.7	0.7	0.7	15.3	8.9	8.9
28	0.0	0.0	0.0	0.0	0.0	0.0
29	0.0	0.0	0.0	0.0	0.0	0.0
30	0.0	0.0	0.0	0.0	0.0	0.0
Avg	15.16	12.26	12.26	35.44	34.74	30.69

Table 13. Passenger delays.

Scen	TPD_3 (min) in the Solution of		
	ALG1	ALG2	Optimal
1	305.7	305.7	305.7
2	2520.8	1955.9	1949.1
3	2725.1	3292.6	2584.8
4	657.9	631.3	627.0
5	1535.9	1431.9	1431.9
6	3959.2	4384.7	3767.0
7	2819.6	2721.4	2721.4
8	449.9	235.1	235.1
9	1641.0	1551.0	721.2
10	1356.4	1255.2	1252.4
11	62.6	62.6	62.6
12	26.6	26.6	26.6
13	380.3	380.3	380.3
14	827.2	1192.9	827.2
15	4216.4	4100.9	3966.0
16	0.0	0.0	0.0
17	217.0	217.0	217.0
18	538.1	1246.4	497.4
19	0.0	0.0	0.0
20	0.0	0.0	0.0
21	142.5	142.5	142.5
22	0.0	0.0	0.0
23	29.7	0.0	0.0
24	260.9	260.9	201.5
25	165.6	165.6	165.6
26	0.0	0.0	0.0
27	258.5	130.9	130.9
28	0.0	0.0	0.0
29	0.0	0.0	0.0
30	0.0	0.0	0.0
Avg	836.56	856.38	740.44

Table 14. Delay propagation.

Secondary Delays	Algorithm	Disturbance Scenario																													
		1	2	3	4	5	6	7	8	9	10	11	12	13	14	15	16	17	18	19	20	21	22	23	24	25	26	27	28	29	30
≤3 min	ALG1	5	5	6	1	4	2	6	16	4	4	4	3	6	1	2	1	2	10	-	2	2	3	3	-	1	3	8	3	3	3
	ALG2	4	7	3	1	3	3	4	8	9	7	3	2	5	1	1	1	2	2	-	3	2	3	-	-	-	-	6	1	-	2
>3 min	ALG1	3	2	9	3	2	7	5	8	8	6	2	1	3	4	4	-	-	1	-	1	-	-	1	-	1	-	2	-	-	-
	ALG2	3	2	3	3	2	7	5	6	8	4	2	1	3	-	4	-	-	4	-	-	-	-	1	-	-	-	-	-	-	-

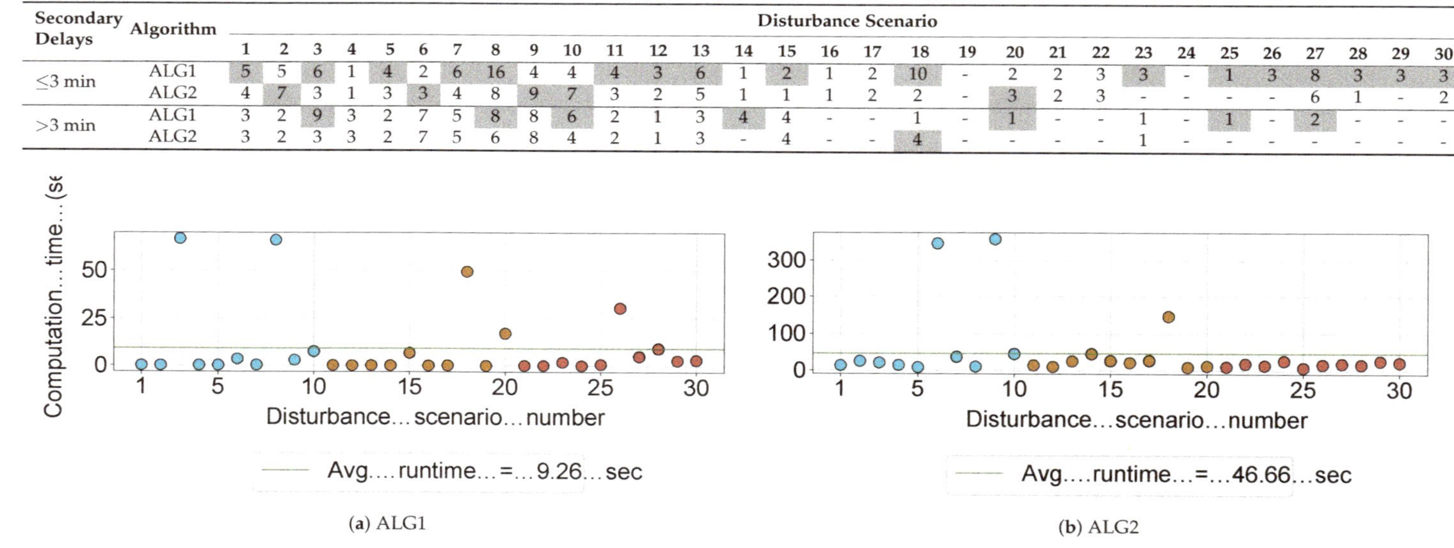

(**a**) ALG1

(**b**) ALG2

Figure 5. Computation times.

Table 15. An overview of train punctuality: average values over all scenarios.

Algorithm	Percentage of Trains					
	Early	On-Time	Delayed at Final Section			
			<3 min	[3, 10) min	[10, 30) min	≥30 min
ALG1	5.10%	85.30%	7.03%	1.43%	0.83%	0.07%
ALG2	0.07%	91.83%	6.03%	1.36%	0.50%	0.07%

Table 16. An overview of train delays: average values over all scenarios.

Algorithm	TFD_3	TAD_3
ALG1	15.16 min	35.44 min
ALG2	12.26 min	34.74 min
Ideal point	12.26 min	30.69 min

Table 17. An overview of delay propagation: average values over all scenarios.

Algorithm	Secondary Delays	
	≤3 min	>3 min
ALG1	3.8	2.4
ALG2	2.8	1.9

Table 18. An overview of freight train performance: average values over all scenarios.

Algorithm	Departure Deviations (d)		Increase in Travel Times (i)	Unplanned Stops (u)	Freight Trains Arriving Earlier than 15 min
	Early	Delayed			
ALG1	2.76 min	0.54 min	1.54 min	2.13	3.63%
ALG2	0.46 min	0.24 min	1.92 min	1.27	0%

Table 19. An overview of track reassignments: average values over all scenarios.

Algorithm	Passenger Trains		Freight Trains		Total
	At Stations	At Lines	At Stations	At Lines	
ALG1	1.37	2.83	0.30	0.70	5.20
ALG2	0.57	0	0	0	0.57

8.1. Train Punctuality

A general observation is that, in the solutions obtained by ALG2, the percentage of trains exactly on time is typically higher compared to the solutions obtained by ALG1 (see Figure 2b). Furthermore, in the solutions obtained using ALG1, trains often reach their final destination earlier than initially planned in the original timetable, while, in the solutions generated by ALG2, trains rarely arrive at their final station earlier than the originally planned arrival time (see Figure 2a). This makes sense as there is no penalty for early train arrivals in ALG1, whereas one of the objectives of ALG2 is to minimize the end time deviations in train events, albeit the objective with the least priority.

What can also be observed from Figure 2b is that the share of affected trains is significantly larger in scenarios 21–30, which primarily is an effect from the initial source of delay, i.e., a temporary infrastructure failure that immediately affects multiple trains.

ALG1 typically provides solutions with a higher percentage of delayed trains (see Table 10). In the solutions to scenarios 1–10 obtained using ALG1, no train experiences a delay $\geq$ 30 min at their final stations. In the solutions obtained by the algorithms for scenarios 21–30, trains are always punctual at their final stations within 10 min (see Table 10). However, in the solutions obtained by ALG1, more trains experience smaller delays. See the comparatively higher percentages of delayed trains for ALG1 for scenarios 21–30 in Table 10.

8.2. Train Delays

According to the average values in Table 16, ALG2 outperforms ALG1 in obtaining solutions with smaller train delays. The statistical significance of the difference in the performance of the two algorithms could be confirmed only for TFD_3 ($p = 0.002$). For TAD_3, statistical significance could not be confirmed from the obtained results ($p = 0.29$). This means that the difference in the values of TAD_3 for the two algorithms is more likely to have occurred by chance.

For several of the scenarios 1–10, the TFD_3 in ALG1's solutions is often rather far from the optimal value (see Table 12). In contrast, ALG2 always obtains a solution with optimal TFD_3, even if it means causing a large delay to a single train. For example, in the solution obtained by ALG2 for scenario 6, although the solution's TFD_3 is minimized (54.9 min), a train experiences a delay $\geq$ 30 min at its final station (see Table 10). ALG1's solution for scenario 6 has a significantly larger TFD_3 (78.1 min). However, in that solution, no train experiences a final delay $\geq$ 30 min. This is a good example of the trade-off between reducing individual train delays and reducing total train delays.

For a majority of the scenarios 11–30, both the algorithms found either ideal solutions or solutions that are very close to ideal (see Table 12). This is an interesting result, as the ideal point is generally expected to be unattainable [33]. A trade-off was expected between minimizing final and accumulated delays; we did not expect to obtain a solution with optimal TFD_3 as well as optimal TAD_3. It is surprising that such a trade-off between TFD_3–TAD_3 does not occur more frequently in Table 12.

An interesting observation can be made from the results obtained for scenarios 3, 6, 14, and 18. For these four scenarios, ALG1 produces solutions that have a smaller TAD_3, compared to ALG2 (see Table 12). Note that ALG1 does not try to minimize TAD_3. However, in the obtained solutions, the delays accumulated by trains at commercial stops are close to optimal. On the other hand, ALG2 has minimizing TAD_3 as its second objective. A reason for this anomaly is that ALG1, while minimizing the passenger delays, indirectly reduces the delays experienced by trains at commercial stops. On the other hand, once ALG2 optimizes the final delays (i.e., the value of TFD_3) and uses it as a bound, it cannot reduce the accumulated delays beyond a certain point.

8.3. Delay Propagation

According to the average delay propagation values in Table 17, ALG2 outperforms ALG1 in obtaining solutions with less delay propagation. We could confirm the difference in the performance of the two algorithms as statistically significant only for secondary delays $\leq$ 3 min ($p = 0.03$). The differences in the obtained secondary delays $>$ 3 min are not statistically significant ($p = 0.09$). The latter shows that the performance difference is likely to have occurred by chance.

In the solutions of ALG1, many trains often incur secondary delays (see Table 14). In comparison, the solutions obtained using ALG2 typically have fewer trains with secondary delays. In the solutions obtained by ALG1 for disturbances 21–30, the delay caused by the disturbance is almost always propagated to other trains. In comparison, in the solutions obtained by ALG2, there is less propagation of delays caused by the disturbance (see Table 14). When using ALG1, secondary delays $>$ 3 min also appear more frequently, compared to ALG2.

8.4. Freight Train Performance

According to the average values of the metrics used for freight train performance (Table 18), the rescheduling strategy used by ALG1 is problematic from a freight train perspective. Note that none of the algorithms explicitly optimize any metric related to freight trains. In addition, note that, in this experiment, there is no additional time associated with enforcing an unplanned train stop, unlike in, e.g., the model adopted in [19]. Hence, no correlation between the increase of unplanned stops and travel times is expected.

The solutions obtained by the algorithms for scenario 7 are interesting; the freight trains have eight unplanned stops (see Figure 3). For this scenario, ALG1 obtained a solution (i) with larger deviation in freight train departure times ($d = 16$ min) compared to ALG2 ($d = 8$ min), (ii) with minimal increase in travel times ($i = 2$ min) compared to ALG2 ($i > 10$ min). Hence, with respect to freight train travel times, the solution of ALG1 may be seen as a good alternative to ALG2's solution.

Disturbance scenarios 11–20 are those where a passenger train runs slower throughout its route. For a majority of these scenarios, ALG2 obtains solutions in which the values of d, i, and u are negligible (see Figure 3). ALG2 shows a similar performance for disturbance scenarios 21–30, wherein it obtains solutions with small values of the considered metrics. In the solutions obtained by ALG1 for the last ten scenarios, the freight trains incur comparatively (i) large departure deviations, (ii) larger number of unplanned stops, and (iii) higher increase in travel times (see Figure 3). The rescheduling performed by ALG1 often caused many freight trains to arrive early (see Figure 4), even when the train initially affected by the disturbance is a passenger train.

8.5. Passenger Delays

The average passenger delays for ALG1 and ALG2, across all the scenarios, are 837 min and 856 min, respectively. This difference in the performance of the two algorithms concerning passenger delays could not be confirmed as statistically significant ($p = 0.68$). This means that the difference in the values of TPD_3 for the two algorithms is more likely to have occurred by chance.

ALG2 often obtained solutions with TPD_3 within 1% of the optimal TPD_3 (see Table 13). For scenarios 3, 6, 14, and 18, ALG1 produces solutions that have a significantly smaller TPD_3, compared to ALG2 (see Table 13). This shows a strength of its approach which simultaneously considers minimizing TPD_3 and TFD_3 with equal priority. On the other hand, ALG2 has minimizing TPD_3 as its third objective. For the aforementioned scenarios, after ALG2 optimizes the train delays and uses them as bounds, it cannot reduce the passenger delays beyond a certain point.

8.6. Track Reassignments

The average track reassignments in the solutions obtained by ALG1 and ALG2, rounded to the nearest integer, are 5 and 1, respectively (see Table 19). This difference in algorithm performance concerning total track reassignments is statistically significant ($p = 0.01$).

In the configuration of ALG2, minimizing the number of track reassignments is an objective that has little priority. Irrespective of that, the algorithm produces solutions with minimal track reassignments. It is reasonable to assume that the optimal number of track reassignments in a rescheduled timetable is zero. In other words, for the input dataset, a rescheduled timetable can be obtained without making any track reassignments. With that in mind, one can say that, over the entire dataset, ALG2 achieved a very good trade-off between minimizing the delays and the number of track reassignments. Thus, ALG2 produces solutions with minimal track reassignments while giving close-to-optimal train and passenger delays.

For scenarios 1–20, in the solutions output by both the algorithms, only passenger trains incur track reassignments. In case of ALG1, the reason for this is as follows. ALG1 does not reallocate the tracks of trains that are not directly affected by the disturbance. In scenarios 1–20, since the initially disturbed train is a passenger train, only the track allocation of that train is modified during rescheduling. A consequence of this rescheduling strategy is as follows. In each of these scenarios, all the track reassignments in the solutions obtained by ALG1 belong to one train, since ALG1 confines the track reassignments to the initially disturbed passenger train. In contrast, ALG2 changes the tracks of various passenger trains at stations. Only one passenger train incurring a reasonable number of track reassignments could be perceived as an advantage of using ALG1, from a passenger perspective as well as from a dispatching perspective. The reason for the latter is that fewer rescheduled trains make it easier for the dispatcher to supervise during a disturbance.

For scenarios 26–30, the solutions obtained using ALG1 involve many track reassignments (see Table 11). The reason is that, when ALG1 encounters a conflict involving a train with primary delay, it first tries to resolve the conflict by reallocating the train's track. Thus, for the trains with a primary delay, the algorithm prefers track reassignment over retiming and reordering. In each of the disturbances 26–30, more than 24 trains incur primary delays (as shown in Table 7). Due to the rescheduling strategy employed by ALG1, the solutions obtained for these scenarios have many track reassignments (see Table 11). In contrast, for these scenarios, the solutions produced by ALG2 do not involve any track reassignments. Both the algorithms obtained almost-ideal train delays and passenger delays in the solutions for disturbances 26–30. Since ALG2 obtained these solutions without performing any track reassignments, it shows that, for these disturbance scenarios, there is no trade-off between minimizing the number of track reassignments and minimizing the values of train and passenger delays.

8.7. Computation Times

On average, ALG1 is five times faster than ALG2. It takes around 9 seconds to reach completion, compared to the latter algorithm's average computation time of 47 seconds (see Figure 5). This difference in performance of the two algorithms with respect to computation times is statistically significant ($p < 0.001$).

ALG1 solves any disturbance scenario in the dataset to completion in about 1 min. ALG2 can take up to 6 min to solve specific scenarios to completion in the dataset. Interesting disturbances occur in scenarios 6 and 9, where ALG2 takes more than 5 min to find the solution and prove its optimality. These two are the scenarios for which the Pareto-optimal solution obtained by ALG2 has a non-optimal TAD_3 and TPD_3 (see Tables 12 and 13). This means that, for these scenarios, a trade-off needs to be made between minimizing e.g., TPD_3 and the primary objective of ALG2, which is TFD_3. Thus, compared to the output Pareto-optimal solution, a solution with a lower value of TPD_3 cannot be obtained without increasing the value of e.g., TFD_3. The longer computation times for these scenarios could be due to ALG2 trying to prove the Pareto-optimality of the obtained solution before reaching completion.

Similar to the case of scenarios 6 and 9, ALG2 takes a longer amount of time to reach completion for scenario 18. This is also a disturbance scenario for which the obtained Pareto-optimal solution has a non-ideal TAD_3 and TPD_3. Figure 6 shows the progress of ALG2 while solving scenario 18. Notice that, for this scenario, most of the time is spent in finding solutions rather than proving optimality of the found solutions. While minimizing TPD_3, around 9 sec is taken by ALG2 to realize that the TPD_3 of the obtained solution cannot be improved further, without increasing the values of the primary and the secondary objectives (TFD_3 and TAD_3, respectively). The gaps in Figure 6 correspond to the time taken for presolving and root relaxation.

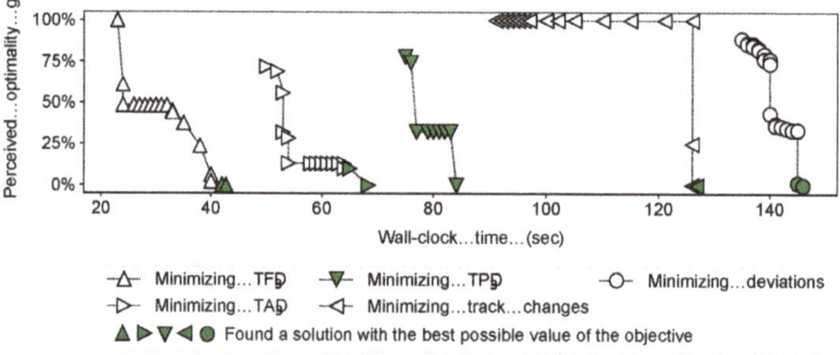

Figure 6. The progress of ALG2 for scenario 18.

From Figure 5, Tables 12 and 13, the following can be observed. For a particular disturbance scenario, whenever there exist solutions in the search space such that there is no significant tradeoff between the first three objectives, ALG2 takes close to the average computation time to find the solution. It is noticed that ALG2 takes longer computation times only when there is a trade-off among the minimization objectives. No such trend could be observed for ALG1, other than the fact that it takes longer than its average time for a few disturbance scenarios.

9. Conclusions and Future Work

The main purpose of the paper was: (i) to present a framework for classification, evaluation and comparison of alternative algorithms and (ii) to demonstrate its application and relevance. The presented framework can be extended and adapted to fit the purpose of other similar evaluation studies. It should be seen as a module-based framework wherein the user can add or exclude certain indicators, e.g., when no freight train traffic is analyzed, the related metrics can be excluded. A user can also include additional metrics of interest [21] in a particular indicator. For example, the maximum secondary delay [42] can be included in the delay propagation indicator. The reason is that it can be useful to log and compare the train that experiences the largest secondary delay as well as the magnitude of that delay.

When using the framework, if a user is satisfied with a subset of the presented indicators and metrics, he/she can choose to stop collecting further measures. Then, as the next step, the user can make a qualitative assessment of particular solutions that each algorithm produces, based on the requirements. An example of such an assessment is shown in [20], where the computed rescheduling solutions are analyzed and scrutinized with other qualitative properties in mind, in addition to the quantitative metrics.

Two noteworthy extensions to the framework are as follows: (i) In addition to measuring the time taken by algorithms to reach completion, one can decide a time limit of e.g., 15 seconds and compare the best solutions obtained by the algorithms within that limit, (ii) For each disturbance scenario, one can collect the solutions obtained by the algorithms during rescheduling as time progresses. The collected solutions can then be analyzed, based on a selected metric, using progress over time plots, such as in [35]. The two extensions are difficult to implement for parallel algorithms, since the order in which solutions are explored/obtained by such algorithms is typically non-deterministic. With large problem datasets, it is not practically viable to implement the second extension into the evaluation framework, even for sequential algorithms.

ALG1 is a heuristic algorithm that considers two minimization objectives with equal priority: minimizing TFD_3 and TPD_3. ALG2 is an exact algorithm that has five minimization objectives, with minimizing TFD_3 as the primary objective. A threshold of 3 min is considered for all the delays appearing in the minimization objectives. Based on the carried out evaluation, we analyze the overall strengths and shortcomings of the two train rescheduling algorithms and their application when solving the 30 disturbance scenarios. A strength of ALG1 is that it is good at quickly finding solutions with small passenger delays. Weaknesses of ALG1 are apparent when it is solving disturbances due to an infrastructure failure. The solutions obtained for these scenarios have significant delay propagation, unsatisfactory freight train performance and many track reassignments.

The strength of ALG2 is its ability to reschedule during infrastructure failures. In the studied scenarios, these failures are of rather modest size. When solving these disturbances, ALG2 is certainly the better choice, since it obtained significantly better solutions and is always within 30 seconds. The main weakness of ALG2 is its speed, particularly while solving disturbances 1–20. Typically, ALG2 obtained good rescheduling solutions for all the considered disturbances. However, compared to ALG1, ALG2 is slow in obtaining solutions. ALG2 took as long as 6 min for a disturbance that is solved in 6 sec using ALG1 (see scenario 6).

When solving disturbances caused by a delayed or a malfunctioned train, a dispatcher can use ALG1 to quickly obtain a decent solution. If the comparatively slower ALG2 is to be used to solve these disturbances, the following are a few suggestions to improve its speed: (i) reduce the number of objectives by merging two or more of the lower-priority objectives into a single objective and (ii) increase the number of parallel threads simultaneously exploring the solution space. A suggestion to improve the practicability of ALG1 is to limit the number of track reassignments while solving disturbances where multiple trains have primary delays. This can result in more practical rescheduling solutions that are easier to implement.

For several of the disturbances considered in the dataset, ideal rescheduling solutions were obtained (with respect to TFD_3, TAD_3 and TPD_3). For most other disturbance scenarios, the Pareto-optimal solution obtained by ALG2 is very close to the hypothetical ideal solution. The frequent existence of a feasible solution in the solution space that simultaneously minimizes TFD_3, TAD_3, and TPD_3 is surprising. Future work could investigate the conditions under which a multi-objective train rescheduling problem contains an ideal solution in its solution space, particularly with respect to the train and passenger delays.

Author Contributions: Conceptualization, S.P.J. and J.T.K.; Methodology, S.P.J. and J.T.K.; Software, S.P.J. and J.T.K.; Validation, S.P.J; Formal analysis, S.P.J.; Investigation, S.P.J.; Resources, S.P.J. and J.T.K.; data curation, J.T.K.; Writing—original draft preparation, S.P.J.; Writing—review and editing, S.P.J., J.T.K. and L.L.; Visualization, S.P.J.; Supervision, J.T.K. and L.L.; Project administration, J.T.K.; Funding acquisition, J.T.K. All authors have read and agreed to the submitted version of the manuscript.

Funding: This research was funded by the Shift2Rail project: FR8Rail II (Grant No: 826206), and the research project BLIXTEN (Grant No: TRV 2018/108023) funded by Trafikverket via KAJT.

Acknowledgments: Many thanks to the anonymous reviewers for their constructive comments, which greatly improved the paper. The authors would also like to thank Tomas Lidén and Martin Joborn for providing valuable comments on several sections of a preliminary version of this paper.

Conflicts of Interest: The authors declare no conflict of interest. The funders had no role in the design of the study; in the collection, analyses, or interpretation of data; in the writing of the manuscript, or in the decision to publish the results.

Abbreviations

The following abbreviations are used in this manuscript:

MILP	Mixed integer linear program
TMS	Traffic management system
IM	Infrastructure manager
TOC	Train operating company
TFD	Total final delay
TAD	Total accumulated delay
TPD	Total passenger delay
DFS	Depth-first search
TOPSIS	Technique for Order of Preference by Similarity to Ideal Solution

References

1. Rao, X.; Montigel, M.; Weidmann, U. A new rail optimisation model by integration of traffic management and train automation. *Transp. Res. Part Emerg. Technol.* **2016**, *71*, 382–405. [CrossRef]
2. Terlaky, T.; Anjos, M.F.; Ahmed, S. *Advances and Trends in Optimization with Engineering Applications*; SIAM: Philadelphia, PA, USA, 2017; Volume 24. [CrossRef]
3. Lamorgese, L.; Mannino, C.; Piacentini, M. Integer Optimization Techniques for Train Dispatching in Mass Transit and Main Line. In *Advances and Trends in Optimization with Engineering Applications*; SIAM: Philadelphia, PA, USA, 2017; pp. 65–75. [CrossRef]
4. Borndörfer, R.; Klug, T.; Lamorgese, L.; Mannino, C.; Reuther, M.; Schlechte, T. Recent success stories on integrated optimization of railway systems. *Transp. Res. Part Emerg. Technol.* **2017**, *74*, 196–211. [CrossRef]
5. Peterson, A.; Wahlborg, M.; Häll, C.H.; Schmidt, C.; Kordnejad, B.; Warg, J.; Johansson, I.; Joborn, M.; Gestrelius, S.; Törnquist Krasemann, J.; et al. Deliverable D 3.1: Analysis of the Gap between Daily Timetable and Operational Traffic. Available online: https://www.semanticscholar.org/paper/Deliverable-D-3.1%3A-Analysis-of-the-gap-between-and-Peterson-Wahlborg/e27a5bff7cead805b824ebce44d2223a7a495b65 (accessed on 26 October 2020).
6. Fang, W.; Yang, S.; Yao, X. A survey on problem models and solution approaches to rescheduling in railway networks. *IEEE Trans. Intell. Transp. Syst.* **2015**, *16*, 2997–3016. [CrossRef]
7. Wegele, S.; Corman, F.; D'Ariano, A. Comparing the effectiveness of two real-time train rescheduling systems in case of perturbed traffic conditions. *Comput. Railw.* **2008**, *103*, 535–544. [CrossRef]
8. Flier, H.F. Optimization of Railway Operations: Algorithms, Complexity, and Models. Ph.D. Thesis, ETH Zurich, Zurich, Switzerland, 2011.
9. Cacchiani, V.; Huisman, D.; Kidd, M.; Kroon, L.; Toth, P.; Veelenturf, L.; Wagenaar, J. An overview of recovery models and algorithms for real-time railway rescheduling. *Transp. Res. Part Methodol.* **2014**, *63*, 15–37. [CrossRef]
10. Morant, A. Dependability and Maintenance Analysis of Railway Signalling Systems. Ph.D. Thesis, Luleå University of Technology, Luleå, Sweden, 2014.
11. Van Thielen, S. Conflict Prevention Strategies for Real-Time Railway Traffic Management. Ph.D. Thesis, KU Leuven, Leuven, Belgium, 2019.

12. Schipper, D.; Gerrits, L. Differences and similarities in European railway disruption management practices. *J. Rail Transp. Plan. Manag.* **2018**, *8*, 42–55. [CrossRef]
13. Tschirner, S. The GMOC Model: Supporting Development of Systems for Human Control. Ph.D. Thesis, Uppsala University, Uppsala, Sweden, 2015.
14. Bettinelli, A.; Santini, A.; Vigo, D. A real-time conflict solution algorithm for the train rescheduling problem. *Transp. Res. Part Methodol.* **2017**, *106*, 237–265. [CrossRef]
15. Törnquist, J. Computer-based decision support for railway traffic scheduling and dispatching: A review of models and algorithms. In Proceedings of the 5th Workshop on Algorithmic Methods and Models for Optimization of Railways (ATMOS'05), Palma, Spain, 14 September 2006.
16. Samà, M.; Meloni, C.; D'Ariano, A.; Corman, F. A multi-criteria decision support methodology for real-time train scheduling. *J. Rail Transp. Plan. Manag.* **2015**, *5*, 146–162. [CrossRef]
17. Fan, B.; Roberts, C.; Weston, P. A comparison of algorithms for minimising delay costs in disturbed railway traffic scenarios. *J. Rail Transp. Plan. Manag.* **2012**, *2*, 23–33. [CrossRef]
18. Min, Y.H.; Park, M.J.; Hong, S.P.; Hong, S.H. An appraisal of a column-generation-based algorithm for centralized train-conflict resolution on a metropolitan railway network. *Transp. Res. Part Methodol.* **2011**, *45*, 409–429. [CrossRef]
19. Törnquist, J.; Persson, J.A. N-tracked railway traffic re-scheduling during disturbances. *Transp. Res. Part Methodol.* **2007**, *41*, 342–362. [CrossRef]
20. Törnquist Krasemann, J. Computational decision-support for railway traffic management and associated configuration challenges: An experimental study. *J. Rail Transp. Plan. Manag.* **2015**, *5*, 95–109. [CrossRef]
21. Corman, F.; Quaglietta, E.; Goverde, R.M. Automated real-time railway traffic control: An experimental analysis of reliability, resilience and robustness. *Transp. Plan. Technol.* **2018**, *41*, 421–447. [CrossRef]
22. Josyula, S.P.; Törnquist Krasemann, J.; Lundberg, L. Parallel computing for multi-objective train rescheduling. *IEEE Trans. Emerg. Top. Comput.* **2020**. [CrossRef]
23. Josyula, S.P.; Törnquist Krasemann, J. Passenger-oriented railway traffic re-scheduling: A review of alternative strategies utilizing passenger flow data. In Proceedings of the 7th International Conference on Railway Operations Modelling and Analysis, Lille, France, 4–7 April 2017.
24. Harrod, S.; Schlechte, T. A direct comparison of physical block occupancy versus timed block occupancy in train timetabling formulations. *Transp. Res. Part Logist. Transp. Rev.* **2013**, *54*, 50–66. [CrossRef]
25. Lamorgese, L.; Mannino, C.; Pacciarelli, D.; Törnquist Krasemann, J. Train Dispatching. In *Handbook of Optimization in the Railway Industry*; Springer: New York, NY, USA, 2018; pp. 265–283. [CrossRef]
26. Corman, F.; Quaglietta, E. Closing the loop in real-time railway control: Framework design and impacts on operations. *Transp. Res. Part Emerg. Technol.* **2015**, *54*, 15–39. [CrossRef]
27. Toletti, A.; De Martinis, V.; Weidmann, U. What about train length and energy efficiency of freight trains in rescheduling models? *Transp. Res. Procedia* **2015**, *10*, 584–594. [CrossRef]
28. Trafikverket Network Statement. 2020. Available online: www.trafikverket.se/en/startpage/operations/Operations-railway/Network-Statement/network-statement-2020/ (accessed on 26 October 2020).
29. Lamorgese, L.; Mannino, C. An Exact Decomposition Approach for the Real-Time Train Dispatching Problem. *Oper. Res.* **2015**, *63*, 48–64. [CrossRef]
30. Li, M.; Yao, X. Quality Evaluation of Solution Sets in Multiobjective Optimisation: A Survey. *ACM Comput. Surv.* **2019**, *52*. [CrossRef]
31. Harrod, S.; Cerreto, F.; Nielsen, O.A. A closed form railway line delay propagation model. *Transp. Res. Part Emerg. Technol.* **2019**, *102*, 189–209. [CrossRef]
32. Hwang, C.L.; Masud, A.S.M. *Multiple Objective Decision Making—Methods and Applications: A State-of-the-Art Survey*; Springer: Berlin/Heidelberg, Germany, 2012.
33. Arora, J.S. Chapter 18—Multi-objective Optimum Design Concepts and Methods. In *Introduction to Optimum Design*, 4th ed.; Academic Press: Boston, MA, USA, 2017; pp. 771–794. [CrossRef]
34. Opricovic, S.; Tzeng, G.H. Compromise solution by MCDM methods: A comparative analysis of VIKOR and TOPSIS. *Eur. J. Oper. Res.* **2004**, *156*, 445–455. [CrossRef]
35. Josyula, S.P.; Krasemann, J.T.; Lundberg, L. A parallel algorithm for train rescheduling. *Transp. Res. Part Emerg. Technol.* **2018**, *95*, 545–569. [CrossRef]

36. Optimization, L.G. Gurobi Optimizer Reference Manual Version 9.0. 2020. Available online: https://www.gurobi.com/wp-content/plugins/hd_documentations/documentation/9.0/refman.pdf (accessed on 26 October 2020).
37. Trafikverket. Maps of Traffic Management Areas. 2017. Available online: https://www.trafikverket.se/for-dig-i-branschen/jarnvag/Trafikledning (accessed on 28 November 2017).
38. Verma, J.; Abdel-Salam, A.S.G. *Testing Statistical Assumptions in Research*; John Wiley & Sons: Hoboken, NJ, USA, 2019.
39. McDonald, J.H. *Handbook of Biological Statistics*; Sparky House Publishing: Baltimore, MD, USA, 2009.
40. Marino, M.J. Chapter 3—Statistical Analysis in Preclinical Biomedical Research. In *Research in the Biomedical Sciences*; Academic Press: Cambridge, MA, USA, 2018; pp. 107–144. [CrossRef]
41. Beiranvand, V.; Hare, W.; Lucet, Y. Best practices for comparing optimization algorithms. *Optim. Eng.* **2017**, *18*, 815–848. [CrossRef]
42. D'ariano, A.; Pacciarelli, D.; Pranzo, M. A branch and bound algorithm for scheduling trains in a railway network. *Eur. J. Oper. Res.* **2007**, *183*, 643–657. [CrossRef]

Article

A Comparison of Ensemble and Dimensionality Reduction DEA Models Based on Entropy Criterion

Parag C. Pendharkar

Information Systems School of Business Administration, Pennsylvania State University, Harrisburg 777 West Harrisburg Pike, Middletown, PA 17057, USA; pxp19@psu.edu; Tel.: +1-717-948-6028

Received: 28 July 2020; Accepted: 14 September 2020; Published: 16 September 2020

Abstract: Dimensionality reduction research in data envelopment analysis (DEA) has focused on subjective approaches to reduce dimensionality. Such approaches are less useful or attractive in practice because a subjective selection of variables introduces bias. A competing unbiased approach would be to use ensemble DEA scores. This paper illustrates that in addition to unbiased evaluations, the ensemble DEA scores result in unique rankings that have high entropy. Under restrictive assumptions, it is also shown that the ensemble DEA scores are normally distributed. Ensemble models do not require any new modifications to existing DEA objective functions or constraints, and when ensemble scores are normally distributed, returns-to-scale hypothesis testing can be carried out using traditional parametric statistical techniques.

Keywords: data envelopment analysis; dimensionality reduction; ensembles; exhaustive state space search; entropy

---

1. Introduction

Data envelopment analysis (DEA) is a prominent technique for the non-parametric relative efficiency analysis of a set of decision-making units (DMUs) drawn from a similar production process [1]. DEA models are used in both operation research and data mining literature [2]. Some of the traditional properties of production functions, such as the monotonicity and convexity of the inputs and outputs, that are fundamental in DEA models are often found to be attractive in some data mining models where datasets are noisy and model resistance to learning noise is necessary [3]. An important aspect of DEA models is the reliability of DMU efficiency scores. It is generally accepted that the DEA efficiency estimates are reliable when the sample size is large [4]. Since the reliability of the DEA scores is dependent on the sample size, Cooper et al. [5] have suggested the following rule for the minimum number (n) of DMUs for reliable DEA analysis (each DMU has m inputs and s outputs):

$$n \geq max\{3(m+s),\ m \times s\} \tag{1a}$$

For small-size datasets, where violations of the minimum number of DMUs specified by Equation (1a) frequently occur, dimensionality reduction (also known as variable reduction or variable selection) approaches are frequently used to select a subset of variables to satisfy Equation (1a). A variety of variable selection approaches are available in the literature. Among these variable selection approaches are statistical [6], regression [7], efficiency contribution measure [8], bootstrapping [9], hypothesis testing [10], variable aggregation [11] and statistical experiment designs [12]. Variable selection approaches are criticized extensively for applying parametric procedures and linear relationship assumptions for selecting variables to determine an unknown non-linear and non-parametric efficiency frontier. Nataraja and Johnson [13] provide a good description of some of these procedures and their pros and cons.

Pendharkar [14] proposed a competing approach to the dimensionality reduction/variable selection problem called the ensemble DEA. In his approach, traditional DEA analysis is conducted for all possible input and output combinations, and the efficiency scores of each DEA model for each DMU are averaged as an ensemble efficiency score for a DMU. Drawing from machine learning literature, Pendharkar [14] showed that the ensemble efficiency score is a reliable estimate of the "true" efficiency of a DMU. Even for small datasets, certain combinations of inputs will satisfy the criterion set by Equation (1a), while others will violate it, but the average ensemble score will be closer to the true efficiency of the DMU and will be reliable. Pendharkar [14] also proposed an exhaustive search procedure to generate all possible input and output combinations, and proposed a formula to compute the number of unique DEA models that need to be run to compute an average ensemble score. This number N of unique DEA models may be computed using the following formula:

$$N = \left(\sum_{i=1}^{m}\binom{m}{i}\right)\times\left(\sum_{i=1}^{s}\binom{s}{i}\right) = (2^m - 1)\times(2^s - 1). \quad (1b)$$

Using Banker et al.'s [15] variable-returns-to-scale (VRS) DEA BCC model, and data and models obtained from a few studies in the literature, Pendharkar [14] showed that the ensemble DEA model provides a better ranking of DMUs than the models proposed in a few studies from the literature.

This research investigates the additional properties and statistical distribution of the ensemble DEA model scores. It is shown that there are added benefits of ensemble efficiency scores. In particular, the ensemble efficiency scores maximize entropy, meaning that the DMU ranking distribution generated by the ensemble efficiency scores has a lower bias when compared to some competing radial and non-radial variable selection models recently reported in the literature, and second, the ensemble efficiency scores may be normally distributed under certain restrictive assumptions. The normal distribution of the efficiency score feature is particularly attractive because returns-to-scale hypothesis testing may be conducted by using traditional difference-in-means parametric statistical procedures. Both of these features are tested using data and models reported in a published study [16]. The rest of the paper is organized as follows: In Section 2, the basic DEA radial and non-radial models, ensemble DEA model and Entropy criterion for comparing different DEA models are described. In Section 3, using Iranian gas company data, the results of ensemble DEA models are compared with the results of variable selection models used in Toloo and Babaee's [16] study. Additionally, in Section 3, the properties of the ensemble DEA scores are investigated in terms of the entropy criterion and their statistical distributions. In Section 4, the paper concludes with a summary and directions for future research.

2. DEA Preliminaries, Ensemble DEA Model, Entropy Criterion for DEA Model Comparisons and Statistical Distribution of Ensemble Scores

The basic DEA model assumes n DMUs, with each DMU consisting of m different inputs that produce s different outputs. The input and output vectors are semi-positive, and for DMU$_j$ ($j = 1, \ldots, n$), the space for the input and output vectors $(x_j, y_j) \in \mathbb{R}_+^{m+s}$. For a DMU$_o$, its relative efficiency may be computed by using the linear programming model under the constant returns-to-scale assumption. This efficiency is computed by solving the following model:

$$\max \sum_{r=1}^{s} u_r y_{ro}, \quad (2a)$$

subject to:

$$\sum_{i=1}^{m} v_i x_{io} = 1 \quad (2b)$$

$$\sum_{r=1}^{s} u_r y_{rj} - \sum_{i=1}^{m} v_i x_{ij} \le 0 \quad for\ all\ j = 1, \ldots, n \quad (2c)$$

$$v_i, u_r \geq \varepsilon \; for \; all \; i = 1, \ldots, m \; and \; r = 1, \ldots, s \tag{2d}$$

where v_i and u_r are the weights associated with the ith input and jth output, respectively. The constant $\varepsilon > 0$ is infinitesimally non-Archimedean. The model (2a)–(2d) is often called the primary CCR model [1], and its dual is written as follows:

$$minimize \; \theta - \varepsilon\left(\sum_{i=1}^{m} s_i^- + \sum_{r=1}^{s} s_r^+\right), \ldots\ldots \tag{2e}$$

subject to:

$$\sum_{j=1}^{n} \lambda_j x_{ij} + s_i^- = \theta x_{io}, \quad i = 1, \ldots, m \tag{2f}$$

$$\sum_{j=1}^{n} \lambda_j y_{rj} - s_r^+ = y_{ro}, \quad r = 1, \ldots, s, \text{ and} \tag{2g}$$

$$\lambda_j, s_i^-, s_r^+ \geq 0 \; for \; all \; i = 1, \ldots, m; j = 1, \ldots, n; r = 1, \ldots, s \tag{2h}$$

The VRS BCC model augments the system (2e)–(2h) by adding the following constraint:

$$\sum_{j=1}^{n} \lambda_j = 1$$

The aforementioned models are radial DEA models that are criticized for not providing input or output projections (for inefficient DMUs) that satisfy Pareto optimality conditions [17]. Fare and Lovell [18] independently proposed radial DEA models that allow for input or output reductions at variable rates. The radial version of the CCR model is mathematically represented in the following dual form:

$$minimize \; \frac{1}{m}\sum_{i=1}^{m} \theta_i$$

subject to:

$$\sum_{j=1}^{n} \lambda_j x_{ij} \leq \theta_i x_{io}, \quad i = 1, \ldots, m$$

$$\sum_{j=1}^{n} \lambda_j y_{rj} \geq y_{ro}, \quad r = 1, \ldots, s$$

$$\theta_i \leq 1, \quad i = 1, \ldots, m \; \ldots\ldots$$

$$\lambda_i \geq 0, \quad j = 1, \ldots, n$$

Pendharkar [14] proposed an ensemble DEA model based on the popularity of ensemble models in machine learning literature. The ensemble DEA model requires an exhaustive search procedure using a binary vector z whose components indicate whether an input or output is considered in performing DEA analysis. The dimension of this binary vector is (m + s). Figure 1 illustrates the z vector and exhaustive search tree for two-input-and-one-output datasets. The exhaustive tree is pruned (dotted edges) for models that have either no inputs or no outputs. DEA analysis is then conducted on the remaining models, and the efficiency results of each model for each DMU are averaged and used as ensemble DEA scores. To illustrate the ensemble DEA approach on a two-input-and-one-output dataset, a CCR DEA analysis using partial Cobb–Douglas production function data on US economic growth between 1899 and 1910 [19] is conducted. Table 1 illustrates the results of our DEA analysis and resulting ensemble scores. The two inputs were labor in person-hours worked per year and the amount of capital invested. The output was the total annual production. The results of the analysis show that the traditional DEA with z = [111] does not provide unique rankings (for the years 1901 and 1902 receive the same efficiency score), but the ensemble DEA model provides unique DMU rankings. Pendharkar's [14] study provides a theoretical basis for the reliability of ensemble DEA scores.

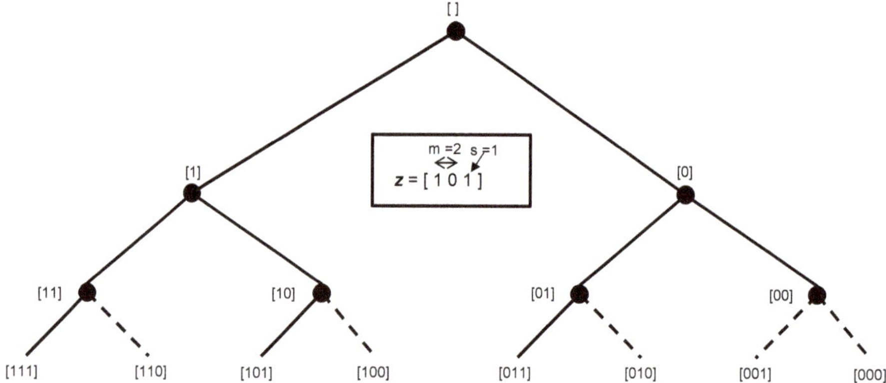

Figure 1. Exhaustive Search Tree for possible unique combinations of two-input-one-output datasets.

Table 1. Ensemble data envelopment analysis (DEA) scores for 1899–1910 US economic growth data.

Year	Production	Labor	Capital	DEA Model Efficiencies			Ensemble Score
				z = [111]	z = [101]	z = [011]	
1899	100	100	100	0.681	0.681	0.665	0.676
1900	101	105	107	0.722	0.722	0.678	0.707
1901	112	110	114	0.693	0.693	0.689	0.692
1902	122	117	122	0.693	0.681	0.693	0.689
1903	124	122	131	0.720	0.720	0.714	0.718
1904	122	121	138	0.770	0.770	0.758	0.766
1905	143	125	149	0.793	0.710	0.793	0.765
1906	152	134	163	0.809	0.730	0.809	0.783
1907	151	140	176	0.836	0.794	0.836	0.822
1908	126	123	185	1.000	1.000	1.000	1.000
1909	155	143	198	0.921	0.870	0.921	0.904
1910	159	147	208	0.941	0.891	0.941	0.924

The maximum entropy (ME) principle has been applied to DEA DMU ranking distribution [20] and model comparisons [21]. The ME principle measures the DMU ranking bias by using a more general family of distributions [22]. Several statistical distributions can be characterized as ME densities [23]. The ME distributions are the least biased distributions obtained by imposing moment constraints that are inherent in the data [21]. To obtain the ME for a given set of DMUs and their efficiencies, normalized ranks are first obtained by computing $\frac{\theta_i^*}{\sum_{i=1}^n \theta_i^*}$, for each DMU, and then computing the ME for a certain model z as follows:

$$ME^z = -\sum_{i=1}^{n}\left(\frac{\theta_i^*}{\sum_{i=1}^n \theta_i^*}\right) ln\left(\frac{\theta_i^*}{\sum_{i=1}^n \theta_i^*}\right)$$

The ME for the DEA models in Table 1 are $ME^{111} = 2.4768$, $ME^{101} = 2.4775$ and $ME^{011} = 2.4757$. The model with labor as an input and production as an output (z = [101]) has the highest entropy and has the least bias, with a maximum difference between DMU efficiencies for closely ranked DMUs for the years 1901 and 1902. The ensemble entropy is 2.4769, and since it is an average of all z-vector combinations, the comparison benchmark for ensemble entropy is the model with z = [111].

The ensemble entropy is higher than the benchmark. The highest possible entropy value or upper bound (UB) for a model is given by the following expression:

$$\mathrm{ME}^{\mathrm{UB}} = -n \times \left(\left(\frac{1}{n}\right) ln\left(\frac{1}{n}\right)\right) \tag{2i}$$

The $\mathrm{ME}^{\mathrm{UB}}$ for the data in Table 1 is 2.485, and the ensemble entropy is very close to the maximum value. It is important to note that obtaining the maximum value is not always desirable, but it provides a theoretical benchmark estimate for a completely unbiased normalized DMU score distribution.

To compute ensemble efficiency scores, an $n \times m$ matrix E of DEA efficiency scores is necessary. The rows of such a matrix are the numbers of DMUs, and the columns are the numbers of models given by the numbers of eligible models considered in computing ensemble efficiency scores. This number of eligible models will have an upper bound given by N, computed using Equation (1b). The elements of this matrix will be efficiency scores for each DMU computed for a given model identified by column number. Figure 2 illustrates a five-DMU-and-five-model matrix. The ensemble efficiency score (θ_i^E) for each DMU is computed using the following formula:

$$\theta_i^E = \frac{\sum_{j=1}^{m} \theta_{ij}^*}{m} \tag{2j}$$

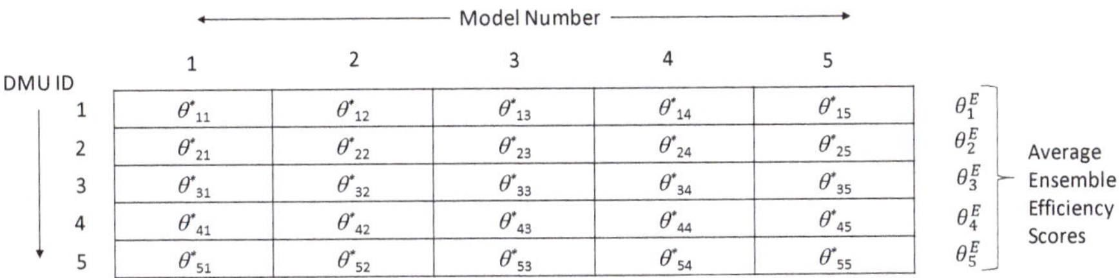

Figure 2. An illustration of 5 × 5 ensemble efficiency score matrix.

A few observations can be made about any row $i \in \{1, \ldots, n\}$ of the ensemble efficiency score matrix. First, all the elements of a given row are an independent computation of efficiency scores by the same DMU under a different model number with its unique set of input(s) and output(s). Second, in all the elements of a given row, the DMU is maximizing its efficiency given its model constraints. Thus, each row represents independent evaluations by a DMU under the maximum decisional efficiency (MDE) principle [24]. The MDE principle was introduced by Troutt [25] to develop a function to the aggregate the performance of multiple decision-makers. The underlying assumption of the MDE principle is that all decision-makers seek to maximize their decisional efficiencies. Troutt [26] later used the MDE approach to rank DMUs and showed that DMUs deemed efficient under MDE are also efficient when ranked using the DEA. For a linear aggregator function, such as the one used in Equation (2j), Troutt [26] illustrated that the decisional efficiencies θ can be described by the following probability density function (pdf):

$$g(\theta) = c_{\alpha} e^{\alpha\theta},\ \alpha > 0 \text{ and } \theta \in [0, 1] \tag{2k}$$

The pdf in (2k) is monotone, increasing on its interval with a mode at $\theta = 1$ (see Figure 5 for illustration). Using the laws of probability, the value of $c_{\alpha} = \alpha\ (e^{\alpha} - 1)^{-1}$. Since each element in a given row of the ensemble efficiency score matrix is an independent evaluation by a decision-maker (i.e., a DMU in an ensemble model) trying to maximize its decisional efficiency θ^*_{ij} for $j = \{1, \ldots, m\}$, the probability density function for each row (DMU) can be written as:

$$g(\theta_i) = c_{\alpha_i} e^{\alpha_i \theta_i},\ \alpha_i > 0 \text{ and } \theta_i \in [0, 1] \tag{2l}$$

The central limit theorem mentions that the cumulative distribution functions (cdfs) of the sums of independently identically distributed random variables asymptotically converge to a Gaussian cdf. The ensemble efficiency scores are normalized sums of independent efficiency assessments that will be distributed with a pdf given by (2l). These sums can be considered independent and identically distributed if $\alpha_1 = \alpha_2 = \ldots = \alpha_n$. Under the restrictive assumption that $\alpha_1 = \alpha_2 = \ldots = \alpha_n$, the ensemble efficiency scores are guaranteed to asymptotically converge to a normal distribution by the central limit theorem. In practice, however, the ensemble efficiency scores are not entirely random or perfectly identically distributed (due to the slight likely variation of Equation (2l)'s α_i parameters for each row), and each ensemble model does introduce a degree of mild randomization. For mild differences in the row pdf parameters α_i, where $\alpha_1 \approx \alpha_2 \approx \ldots \approx \alpha_n$, the ensemble efficiency scores are likely to be normally distributed. A reader may note that under ideal conditions, where $\alpha_1 = \alpha_2 = \ldots = \alpha_n$ and individual DMU scores follow Equation (2l)'s distribution, the entropy of the ensemble scores will be highest and close to the highest upper bound mentioned in Equation (2i) because the distribution in Equation (2i) has a mode of 1 (see Figure 5). Thus, it may be argued that the likelihood of normality of the ensemble scores increases when the entropy of the ensemble scores is closer to its upper bound given by Equation (2i). It is important to note that an entropy equal to the exact value of the upper bound given by Equation (2i) is undesirable because at that value, the distribution is a uniform distribution where all the DMUs are fully efficient for all the models. The entropy of the pdf in Equation (2k) is maximized on the interval [0, 1] when the mean of the distribution is greater than 0.5 [27]. Additionally, another important aspect of the distribution of the ensemble efficiency scores is that both the rows and columns of ensemble efficiency scores (Figure 2) play a role in the pdf of the ensemble efficiency scores because the rows represent sampling from the MDE distributions and the columns represent sampling from the distribution of the sums of independent variables. Larger sample sizes increase the statistical reliability and robustness of the results.

3. Comparing Variable Selection Models and Ensemble Model Using Gas Company Data and Entropy Criterion

For small datasets, many input or output variables are aggregated so that the selected variables satisfy the heuristic given in Equation (1a). There are two problems with all the variable selection approaches. First, they use an artificial criterion to select variables for a non-linear and non-parametric approach. Any artificial/subjective criterion will make some assumptions that are harder to justify. Second, these techniques have several selection parameters and thresholds that often lead to inconsistencies in applying these techniques. For example, Toloo and Babaee [16] illustrate three problems with a variable selection approach and suggested an improved approach. By contrast, the ensemble DEA approach does not make any assumptions, and for small datasets, trying out different input and output combinations and aggregating efficiency scores provide more reliable efficiency estimates than variable selection models. Part of the reason for the stability of ensemble DEA efficiency scores is that, even for small datasets, some DEA models in an ensemble will always satisfy the heuristic given in Equation (1a), which will increase the reliability of the ensemble efficiency scores due to model averaging. This stability of ensemble efficiency scores is illustrated by comparing ensemble scores with the results of models from Toloo and Babaee's [16] study and using the entropy criterion.

To compare the results, the dataset from Toloo and Babaee's [16] study is used. The dataset consists of three inputs and four outputs from an Iranian gas company. The inputs are budget (x_1), staff (x_2) and cost (x_3). The outputs are customers (y_1), the length of the gas network (y_2), the volume delivered (y_3) and gas sales (y_4). Table 2 lists these data. Table 3 lists the efficiency scores of the ensemble DEA with the CCR and BCC models and models used by Toloo and Babaee [16]. Using formula (1b), a total of 105 unique DEA models were used to compute the DEA ensemble efficiency score.

Table 2. The Iranian gas company dataset.

DMU	x_1	x_2	x_3	y_1	y_2	y_3	y_4
1	177,430	401	528,325	801	41,675	77,564	201,529
2	221,338	1094	1,186,905	803	34,960	44,136	840,446
3	267,806	1079	1,323,325	251	24,461	27,690	832,616
4	160,912	444	648,685	816	23,744	45,882	251,770
5	177,214	801	909,539	654	36,409	72,676	443,507
6	146,325	686	545,115	177	18,000	19,839	341,585
7	195,138	687	790,348	695	31,221	40,154	233,822
8	108,146	152	236,722	606	23,889	37,770	118,943
9	165,663	494	523,899	652	25,163	28,402	179,315
10	195,728	503	428,566	959	43,440	63,701	195,303
11	87,050	343	298,696	221	9689	17,334	106,037
12	124,313	129	198,598	565	21,032	30,242	61,836
13	67,545	117	131,649	152	10,398	14,139	46,233
14	47,208	165	228,730	211	9391	13,505	42,094

Table 3. The results of experiments.

DMU	Ensemble CCR	Ensemble BCC	Non-Radial [&]	Radial [&]
1	0.87 (0.15)	0.95 (0.11)	0.98	0.75
2	0.75 (0.30)	0.77 (0.28)	1	1
3	0.61 (0.36)	0.62 (0.36)	0.9	0.82
4	0.71 (0.19)	0.8 (0.19)	0.79	0.63
5	0.77 (0.22)	0.82 (0.21)	0.95	0.83
6	0.58 (0.27)	0.64 (0.27)	0.76	0.64
7	0.54 (0.16)	0.57 (0.14)	0.57	0.47
8	0.98 (0.08)	0.99 (0.04)	1	1
9	0.57 (0.14)	0.6 (0.14)	0.61	0.46
10	0.86 (0.18)	0.96 (0.11)	0.85	0.77
11	0.47 (0.12)	0.63 (0.14)	0.55	0.46
12	0.93 (0.15)	0.96 (0.11)	1	1
13	0.63 (0.13)	0.96 (0.09)	0.68	0.51
14	0.6 (0.24)	0.86 (0.17)	0.56	0.51

[&] Results taken from Toloo and Babaee's [16] study.

The entropies of the Ensemble CCR, Ensemble BCC, Non-Radial and Radial models were 2.616, 2.621, 2.615 and 2.599, respectively. The ME^{UB} from Equation (2i) is 2.639. Comparing the Ensemble CCR with the Non-Radial and Radial CCR models shows that the Ensemble CCR model has a higher entropy. Only the VRS Ensemble BCC model has a higher entropy than the Ensemble CCR model. The standard deviations of the Ensemble BCC model are mostly lower than the CCR model's as well. More importantly, the Ensemble CCR model generates unique rankings for the DMUs, whereas the Non-Radial and Radial models generate a tie for three DMUs. The Ensemble BCC model also generates unique rankings, but the differences occur at the third decimal place. The Ensemble BCC efficiency scores for DMU 10, 12 and 13 were 0.960, 0.959 and 0.962, respectively.

Figures 3 and 4 illustrate the numbers of models (out of 105 total models) where a DMU was fully efficient. These figures are useful for understanding to what extent the assumption $\alpha_1 \approx \alpha_2 \approx \ldots \approx \alpha_n$ was satisfied for the theoretical normal distribution of the ensemble efficiency scores. For these parameters to be similar, the expectation is that a similar number of fully efficient DMUs should exist across all models. Clearly, some DMUs are never fully efficient under any of 105 models and the assumption of identical distributions is violated. While the assumption is violated, Figure 4 illustrates that some DMUs, e.g., 1, 8, 10, 12 and 13, have a somewhat similar number of fully efficient DMUs to others. These ensemble scores of these DMUs may be considered as normalized random sums generated from identical distributions (such as Distribution 1). All of these DMUs have ensemble efficiency scores greater than 0.95. Similarly, DMUs 5, 6 and 11, in Figure 4, have no fully efficient scores, and these may also be considered as random normalized sums generated from identically distributed pdfs (such as Distribution 2).

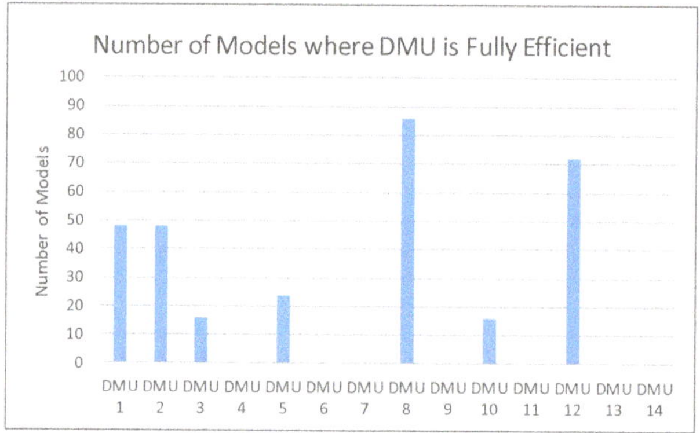

Figure 3. Number of times a DMU is fully efficient in Ensemble CCR models.

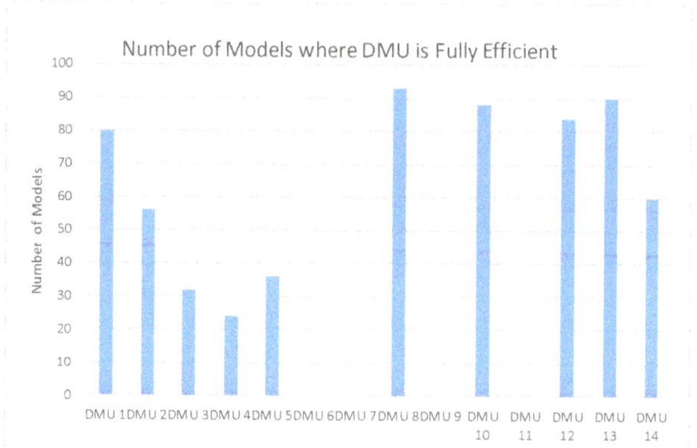

Figure 4. Number of times a DMU is fully efficient in Ensemble BCC models.

The ensemble scores for this dataset appear to be random normalized sums from two or more pdfs of the forms given in Equation (2k). Given that these are independent random normalized sums, it can be easily shown that the product of two or more independent MDE pdfs is also an MDE pdf. Figure 5 illustrates two sample MDE pdfs for two different values of alpha. The entropy of an MDE pdf is maximized when the mean of a distribution is greater than 0.5 [27]. For the ensemble BCC model, from Table 3, this criterion is satisfied. The lowest value of the ensemble BCC score is 0.57, which is greater than the mean of 0.5 required to maximize entropy and higher than the lowest values for the efficiency scores for the radial, non-radial and ensemble CCR models. As a result, the ensemble BCC model appears to maximize its entropy slightly better than the ensemble CCR model.

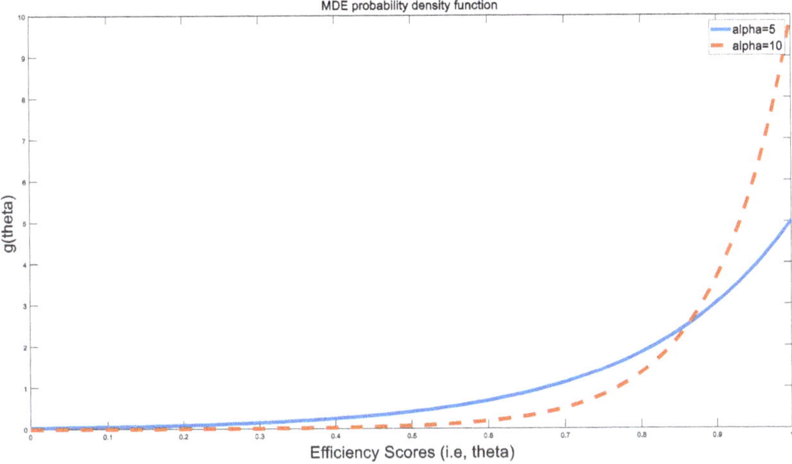

Figure 5. The maximum decisional efficiency (MDE) probability density function (pdf) for $\alpha = 5$ and $\alpha = 10$, respectively.

While ensemble scores have a minor violation of an identical distribution for some DMUs, a formal test of the normality of the distribution of the ensemble efficiency scores was conducted. Table 4 illustrates the results of these tests. The Shapiro–Wilk statistic for the Ensemble CCR model is 0.944, and that for the Ensemble BCC model is 0.876, which, at 14 degrees of freedom, are non-significant, consistent the null hypothesis that the efficiency score distribution is normally distributed at the 95% level of statistical significance.

Table 4. The results of normality tests.

	Kolmogorov–Smirnov			Shapiro–Wilk		
	Statistic	***df***	**Sig.**	**Statistic**	***df***	**Sig.**
Ensemble BCC	0.196	14	0.149	0.876	14	0.051
Ensemble CCR	0.182	14	0.200	0.944	14	0.477

A paired sample t-test for the difference in mean efficiency scores between the Ensemble CCR and the Ensemble BCC models gives a $|t|$-value of 3.524, which is significant at the 99% level of statistical significance ($df = 13$), indicating that a variable returns-to-scale relationship exists between inputs and outputs. The normality of the ensemble efficiency score distributions increases the power of parametric statistical tests.

4. Summary, Conclusions and Directions for Future Work

A significant amount of research in the DEA literature has focused on dimensionality reduction/variable selection techniques for small datasets. These techniques are often criticized and have their limitations, with no clear way of selecting which technique is the best. A better approach would be to use an ensemble DEA score that does not make any additional assumptions and provides models that have high entropy values and normally distributed scores under restrictive assumptions. Pendharkar [14], in his study, has already provided a theoretical foundation for the reliability of ensemble DEA scores. The added benefit of ensemble DEA scores is that they provide unique DMU rankings.

The normality of ensemble DEA scores is not guaranteed unless the ensemble DEA scores are normalized sums generated from independent identically distributed MDE pdfs. This assumption may not be strictly satisfied in most real-world datasets, but the current study shows that minor deviation from this assumption may be tolerated because the entropy of all MDE pdfs is maximized when normalized sums have a value greater than 0.5. This means that, typically, the differences in means between the underlying pdfs (Equation (21)) for ensemble entropy scores will be less than 0.5, and, while these pdfs may not be identically distributed, the means of these distributions will be close, resulting in the likely normal distribution of ensemble scores in most real-world cases. The normality of ensemble DEA scores allows for the application of traditional statistical tests for return-of-scales hypothesis tests. Traditional DEA hypothesis-testing methods are not perfect and are known to be slightly biased [28]. Future research may focus on comparing ensemble DEA-based hypothesis testing with traditional DEA hypothesis testing to identify which method provides reliable results.

Funding: This research received no external funding.

Conflicts of Interest: The authors declare no conflict of interest.

References

1. Charnes, A.; Cooper, W.W.; Rhodes, E. Measuring the efficiency of decision making units. *Eur. J. Oper. Res.* **1978**, *2*, 429–444. [CrossRef]
2. Pendharkar, P.C. Data envelopment analysis models for probabilistic classification. *Comput. Ind. Eng.* **2018**, *119*, 181–192. [CrossRef]
3. Pendharkar, P. A data envelopment analysis-based approach for data preprocessing. *IEEE Trans. Knowl. Data Eng.* **2005**, *17*, 1379–1388. [CrossRef]
4. Ruggiero, J. A new approach for technical efficiency estimation in multiple output production. *Eur. J. Oper. Res.* **1998**, *111*, 369–380. [CrossRef]
5. Cooper, W.W.; Seiford, L.M.; Tone, K. *Data Envelopment Analysis: A Comprehensive Text with Models, Applications, References and DEA-Solver Software*, 2nd ed.; Springer: Berlin/Heidelberg, Germany, 2007.
6. Ueda, T.; Hoshiai, Y. Application of principal component analysis for parsimonious summarization of DEA inputs and/or outputs. *J. Oper. Res. Soc. Jpn.* **1997**, *40*, 466–478. [CrossRef]
7. Ruggiero, J. Impact assessment of input omisson on DEA. *J. Inf. Technol. Decis. Mak.* **2005**, *4*, 359–368. [CrossRef]
8. Wagner, J.M.; Shimshak, D.G. Stepwise selection of variables in data envelopment analysis: Procedures and managerial perspectives. *Eur. J. Oper. Res.* **2007**, *180*, 57–67. [CrossRef]
9. Simar, L.; Wilson, P.W. Testing restrictions in nonparameteric efficiency models. *Commun. Stat.* **2001**, *30*, 159–184. [CrossRef]
10. Banker, R.D. Hypothesis tests using data envelopment analysis. *J. Prod. Anal.* **1996**, *7*, 139–159. [CrossRef]
11. Amirteimoori, A.R.; Despotis, D.K.; Kordrostami, S. Variable reduction in data envelopment analysis. *Optimization* **2012**, *63*, 735–745. [CrossRef]
12. Morita, H.; Avkiran, N.K. Selecting inputs and outputs in data envelopment analysis by designing statistical experiments. *J. Oper. Res. Soc. Jpn.* **2009**, *52*, 163–173. [CrossRef]
13. Nataraja, N.R.; Johnson, A.L. Guidelines for using variable selection techniques in data envelopment analysis. *Eur. J. Oper. Res.* **2011**, *215*, 662–669. [CrossRef]
14. Pendharkar, P.C. Ensemble based ranking of decision making units. *Inf. Syst. Oper. Res.* **2013**, *51*, 151–159. [CrossRef]
15. Banker, R.D.; Charnes, A.; Cooper, W.W. Some models for estimating technical and scale inefficiencies in data envelopment analysis. *Manag. Sci.* **1984**, *30*, 1078–1092. [CrossRef]
16. Toloo, M.; Babaee, S. On variable reductions in data envelopment analysis with an illustrative application to a gas company. *Appl. Math. Comput.* **2015**, *270*, 527–533. [CrossRef]
17. Ray, S.C. *Data Envelopment Analysis: Theory and Techniques for Economics and Operations Research*; Cambridge University Press: Cambridge, UK, 2004.
18. Fare, R.; Lovell, C.A.K. Measuring the technical efficiency. *J. Econ. Theory* **1978**, *19*, 150–162. [CrossRef]
19. Stewart, J. *Multivariable Calculus: Concepts and Contexts*, 3rd ed.; Thomson Learning: London, UK, 2005.
20. Soleimani-Damaneh, M.; Zarepisheh, M. Shannon's entropy for combining the efficiency results of different DEA models: Method and Application. *Expert Syst. Appl.* **2009**, *36*, 5146–5150. [CrossRef]

21. Xie, Q.; Dai, Q.; Li, Y.; Jiang, A. Increasing the discriminatory power of DEA Using Shannon's Entropy. *Entropy* **2014**, *16*, 1571–1585. [CrossRef]
22. Park, S.Y.; Bera, A.K. Maximum entropy autoregressive conditional heteroskedasticity model. *J. Econom.* **2009**, *150*, 219–230. [CrossRef]
23. Kagan, A.M.; Linik, Y.V.; Rao, C.R. *Characterization Problems in Mathematical Statistics*; Wiley: New York, NY, USA, 1973.
24. Pendharkar, P.C. Cross efficiency evaluation of decision-making units using the maximum decisional efficiency principle. *Comput. Ind. Eng.* **2020**, *145*, 106550. [CrossRef]
25. Troutt, M.D. Maximum decisional efficiency estimation principle. *Manag. Sci.* **1995**, *41*, 76–82. [CrossRef]
26. Troutt, M.D. Derivation of the maximin efficiency ratio model from the maximum decisional efficiency principle. *Ann. Oper. Res.* **1997**, *73*, 323–338. [CrossRef]
27. Troutt, M.D.; Zhang, A.; Tadisina, S.K.; Rai, A. Total factor efficiency/productivity ratio fitting as an alternative to regression and canonical correlation models for performance data. *Ann. Oper. Res.* **1997**, *74*, 289–304. [CrossRef]
28. Banker, R.D. Maximum likelihood, consistency and data envelopment analysis: A statistical foundation. *Manag. Sci.* **1993**, *39*, 1265–1273. [CrossRef]

Article

An Unknown Radar Emitter Identification Method Based on Semi-Supervised and Transfer Learning

Yuntian Feng *, Guoliang Wang, Zhipeng Liu, Runming Feng, Xiang Chen and Ning Tai

State Key Lab. of Complex Electromagnetic Environment Effects on Electronics and Information System; Luoyang 471003, China; guoliangwangcemee@163.com (G.W.); zhipengliucemee@163.com (Z.L.); runmingfengcemee@163.com (R.F.); xiangchencemee@163.com (X.C.); ningtaicemee@163.com (N.T.)

* Correspondence: fengyuntian2009@live.cn; Tel.: +86-1873-906-7817

Received: 13 November 2019; Accepted: 10 December 2019; Published: 16 December 2019

Abstract: Aiming at the current problem that it is difficult to deal with an unknown radar emitter in the radar emitter identification process, we propose an unknown radar emitter identification method based on semi-supervised and transfer learning. Firstly, we construct the support vector machine (SVM) model based on transfer learning, using the information of labeled samples in the source domain to train in the target domain, which can solve the problem that the training data and the testing data do not satisfy the same-distribution hypothesis. Then, we design a semi-supervised co-training algorithm using the information of unlabeled samples to enhance the training effect, which can solve the problem that insufficient labeled data results in inadequate training of the classifier. Finally, we combine the transfer learning method with the semi-supervised learning method for the unknown radar emitter identification task. Simulation experiments show that the proposed method can effectively identify an unknown radar emitter and still maintain high identification accuracy within a certain measurement error range.

Keywords: semi-supervised learning; transfer learning; radar emitter

1. Introduction

Radar emitter identification is the key link in radar reconnaissance. It extracts the characteristic parameters and working parameters on the basis of radar signal sorting. Based on these parameters, we can obtain the information such as the system, use, type and platform of the target radar, and further deduce the battlefield situation, threat level, activity rule, tactical intention, etc., and provide important intelligence support for one's own decision-making [1]. The most commonly used radar emitter identification method is the pulse described word-based method. As new radar systems are born, and the radar is becoming more complex, the method is difficult to cope with the complex electromagnetic environment of modern battlefields. In order to obtain better identification results, researchers began to extract a variety of new features in the time domain [2], frequency domain [3] and time-frequency domain [4] for the identification of radar emitters.

With the rise of deep learning techniques, more and more researchers have applied CNN and DBN in the radar emitter identification task, which achieves good performance. Zhou Z et al. [5] developed a novel deep architecture for automatic waveform recognition, which outperformed the existing shallow algorithms and other hand-crafted, feature-based methods. Cain L et al. [6] investigated an application of convolutional neural networks (CNN) for rapid and accurate classification of electronic warfare emitters. Sun J et al. [7] proposed a deep learning model named as unidimensional convolutional neural network (U-CNN) to classify the encoded high-dimension sequences with big data.

Kong M et al. [8] used the CNN deep learning algorithm to identify the radar radiation sources, which could extract more detailed features of the radar and improve the recognition rate. To cope with the complex electromagnetic environment and varied signal styles, Wang X et al. [9] proposed a novel method based on the energy cumulant of short time Fourier transform and reinforced deep belief network to gain a higher correct recognition rate for radar emitter intra-pulse signals at a low signal-to-noise ratio.

In the past battlefields, the types of radar emitters are single and limited, and the above methods can solve the problem of radar emitter identification well. However, with the increasing number and variety of radar emitters, many unknown emitters will appear in the future battlefield. As time goes by and the location changes, the current identification methods will face two problems. First, the training data and testing data no longer satisfy the same-distribution hypothesis, resulting in a decrease in the classification performance of the machine learning model. Second, the number of available labeled samples for unknown emitters is seriously insufficient, which may lead to over-fitting of the machine learning model.

In recent years, the transfer learning methods [10] and the semi-supervised learning methods [11] have gained more and more attention. Transfer learning does not require that the training data and testing data meet the conditions of the same distribution in the model training process, and utilizes the knowledge in a large number of known samples for training, which is good for cross-domain learning. However, the transferring of a large amount of irrelevant information will also cause negative transfer, which reduces the effect of identification. Semi-supervised learning can use the information in a small number of labeled samples and find patterns from a large number of unlabeled samples, and then perform classification, avoiding the use of only a small number of labeled samples for training, which may result in over-fitting. However, as information continues to increase, the training data and testing data will also not satisfy the same-distribution hypothesis.

In view of the different characteristics of transfer learning and semi-supervised learning, this paper combines the two methods to propose an unknown radar emitter identification method based on semi-supervised and transfer learning. Firstly, we construct the support vector machine model based on transfer learning, using the information of labeled samples in the source domain to train in the target domain, which can solve the problem that the training data and the testing data do not satisfy the same-distribution hypothesis. Then we design a semi-supervised, co-training algorithm, using the information of unlabeled samples to enhance the training effect, which can solve the problem that insufficient labeled data results in inadequate training of the classifier. Finally, we combine the transfer learning method with the semi-supervised learning method for the unknown radar emitter identification task.

Our major contributions are summarized as follows: (1) Focusing on the actual application scenarios to study radar emitter identification, and simultaneously solving the problem that training data and testing data do not satisfy the same-distribution hypothesis and the problem of insufficient labeled data, which provides a good thinking way for future research in this area; (2) proposing a method combining support vector machine based on transfer learning with semi-supervised co-training algorithm; (3) verifying the interaction between the transfer learning method and the semi-supervised learning method for unknown radar emitter identification task.

2. Relevant Research

2.1. Transfer Learning

Transfer learning refers to learning the knowledge in the source domain D_s, and using in the target domain D_t that is not the same distribution with D_s but is related to D_s, which makes good the problem of insufficient training data. Unlike traditional machine learning methods, transfer learning [12] does not require training data and testing data to satisfy the same-distribution hypothesis. It can discover and extract knowledge in the source domain D_s that matches the distribution of the target domain D_t and is useful for identification in the target domain D_t.

Then it establishes classification models in the target domain D_t, which can make efficient use of existing labeled samples to avoid re-labeling in the target domain D_t.

From the perspective of transfer methods, transfer learning includes four basic methods: sample-based transfer [13], feature-based transfer [14], model-based transfer [15] and relationship-based transfer [16]. The sample-based transfer method refers to producing rules according to certain weights, and reusing data samples for transfer learning. The feature-based transfer method refers to mutual transfer by feature transformation, which reduces the gap between the source domain and the target domain, or transforms the data features of the source domain and the target domain into a unified feature space, and then utilizes the traditional machine learning methods for identification. The model-based transfer method refers to finding the parameter information shared between the source domain and the target domain to implement transferring. The relationship-based transfer method has a completely different approach from the above three methods, focusing on the similarity between the source domain samples and target domain samples.

2.2. Semi-Supervised Learning

The commonly used machine learning methods can be divided into three categories: supervised learning, unsupervised learning and semi-supervised learning. Supervised learning refers to only using labeled samples for training, and may not obtain a model with high generalization ability in the case of fewer labeled samples. Unsupervised learning refers to only using unlabeled samples for training, regardless of labeled samples, which results in a waste of samples. Semi-supervised learning can process a small number of labeled samples and a large number of unlabeled samples at the same time, combining the advantages of supervised learning and unsupervised learning.

The four most commonly used algorithms for semi-supervised learning are Self-Training [17], Co-Training [18], Generative Model [19] and Graph-Based Semi-supervised [20]. The Self-Training algorithm refers to the use of a self-classifier to continuously generate high-confidence samples for improving the final classification performance. The Co-Training algorithm refers to separately training the classifier on two views, which is representative of multi-view learning. The Generative Model-based method means that the data of different categories meets different distributions, and if its conditional probability distribution is known, the parameters of the model can be solved. The Graph-Based, Semi-supervised method refers to passing the label information of labeled samples to unlabeled samples according to the adjacency relationship in the graph, thereby realizing the classification of the unlabeled samples.

3. Unknown Radar Emitter Identification Based on Semi-Supervised and Transfer Learning

In this section, we use the support vector machine model as the base classifier. Firstly, we construct the support vector machine based on transfer learning, and define the calculation index to measure the transfer ability. Then we study the training effect enhancement method based on the semi-supervised co-training algorithm. Finally, we combine the transfer learning method with the semi-supervised learning method for the unknown radar emitter identification task.

3.1. Support Vector Machine Based on Transfer Learning

The support vector machine (SVM) model has the characteristics of simple structure and global optimization, and is good at solving small sample and nonlinear problems. Therefore, this section chooses the support vector machine model as the base classifier to perform radar emitter identification.

In the process of constructing the SVM model based on transfer learning, it is necessary to utilize the data in two domains at the same time, namely source domain D_s and target domain D_t. The data in source domain D_s refers to the known radar emitters that are detected during non-wartime, and the data in target domain D_t refers to the emerging radar emitters in wartime.

When the amount of data in source domain D_s is large, noise in source domain D_s affects the use of the data in target domain D_t.

In order to better optimize the target equation, this section filters the data with high similarity in the source domain D_s in the process of transferring the SVM model, and uses the Euclidean distance to define the distance function $\sigma(V_s^i, D_t^j)$, which can measure the similarity between the source domain data and the target domain data. Its formula is as follows:

$$\sigma(V_s^i, D_t^j) = -\frac{1}{k} \sum_{(x_j, y_j) \in D_t} \exp\{-\beta \parallel V_s^i - x_j \parallel_2^2\} \tag{1}$$

where V_s^i is the support vector for source domain, β is the importance degree of V_s, (x_j, y_j) is the sample in target domain D_t and its real category, $\parallel V_s^i - x_j \parallel_2^2$ is the Euclidean distance between V_s and D_t, k is the number of samples in target domain D_t.

The specific steps of the support vector machine based on transfer learning are shown in Algorithm 1.

Algorithm 1. Support vector machine based on transfer learning

1. Train the initial SVM model in the source domain D_s to get the support vector V_s, and calculate the similarity distance function $\sigma(V_s^i, D_t^j)$.
2. Add V_s to the source domain data, and add the similarity distance function σ to the objective function of the SVM model as follows:

   $$\min_w 0.5 \parallel w \parallel_2^2 + C \sum_{j=1}^{k} \varepsilon_j + \sum_{i=1}^{m} \sigma(V_s^i, D_t^j) \overline{\varepsilon}_i$$

   Where m is the number of samples in source domain D_s, C is the penalty term, w is the weights of classification hyperplane in the SVM model.
3. Generate new training set $\widetilde{D}$ in target domain D_t, and retrain the SVM model. The optimization problem of the objective function is described with the Lagrangian coefficient as follows:

   $$\max L(\alpha) = \sum_{i=1}^{m+k} \alpha_i - 0.5 \sum_{i=1}^{m+k} \sum_{j=1}^{m+k} \alpha_i \alpha_j y_i y_j (x_i * x_j)$$

   Where $x_i * x_j$ is the inner product of the vector x_i and the vector x_j, y_i is the real category label of x_i, y_j is the real category label of x_j, α_i and α_j are the Lagrangian multipliers, $\alpha = (\alpha_1, \alpha_2, \cdots, \alpha_{m+k})^T$ is the Lagrangian multiplier vector.
4. Solve the above optimization problem and get the optimal solution α^*, which means getting the final SVM model. Its form is as follows:

   $$f(x) = sign[\sum_{j=1}^{m+k} y_i \alpha^* (x_i * x_j) + y_i - \varepsilon_i - (\sum_{i=1}^{m+k} \alpha^* x_i y_i)^T x_i]$$

3.2. Transfer Ability

The transfer ability can reflect the influencing ability of the samples in source domain D_s on the target domain D_t. The calculation process involves two important indices: the similarity between the sample in source domain D_s and the sample in target domain D_t; the consistency between the prediction result of the sample x_i in the classifier f and its real category. Therefore, the calculation formula of transfer ability is as follows:

$$\alpha_i = \sigma(V_s^i, D_t^j) * f(x_i) == y_i \tag{2}$$

where $\sigma(V_s^i, D_t^j)$ is the similarity distance function, f is the SVM classifier trained by the above transfer learning method, $f(x_i)$ is the predicted value of x_i in source domain D_s by the classifier f, y_i is the real category label of x_i. By calculating the transfer ability, it is helpful to select the samples in source domain D_s which are related to target domain D_t.

3.3. Training Effect Enhancement Based on Semi-Supervised Co-Training Algorithm

The transfer learning-based support vector machine can select the appropriate samples from source domain D_s for the training on target domain D_t, which can improve the final identification performance. Unlike the above, semi-supervised learning can use the unlabeled samples in target domain D_t to enhance the final training effect. This section constructs the semi-supervised co-training algorithm based on the base classifier SVM model. The specific steps are shown in Algorithm 2.

Algorithm 2. Semi-supervised co-training algorithm

1. For the radar emitter identification task, define and construct a feature set $\mathbf{x}$, and divide it into two parts $\mathbf{x}_1$ and $\mathbf{x}_2$.
2. Train the base classifier SVM model on the small number of labeled samples in target domain D_t by using the feature sets $\mathbf{x}_1$ and $\mathbf{x}_2$ respectively, and obtain the classifiers f_1 and f_2.
3. For t = 1: N

   Perform identification on the unlabeled samples in target domain D_t by using the classifiers f_1 and f_2, respectively, and obtain the posterior probabilities of the samples belonging to each emitter category, and select p samples with the highest confidence for each category;
   Add the selected samples to the training set and retrain the classifiers f_1 and f_2 on the training set.

End

The two feature sets $\mathbf{x}_1$ and $\mathbf{x}_2$ in the co-training algorithm refer to two views and need to satisfy sufficient redundancy and conditional independence. Through continuous iterative training, unlabeled samples in the target domain D_t are available for labeling, which helps to enhance the training effect.

3.4. Combination of Transfer Learning Method and Semi-Supervised Learning Method

This section combines the transfer learning method with the semi-supervised learning method, while taking advantage of the two methods, which can use useful information in source domain D_s for cross-domain learning, and can enhance the training effect with unlabeled samples in target domain D_t. The specific process is shown in Figure 1.

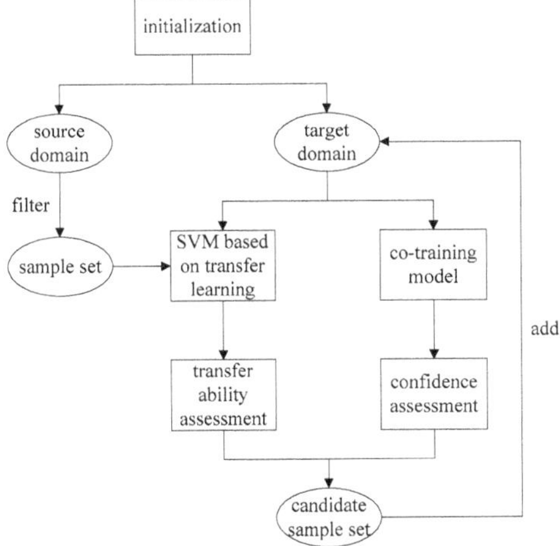

Figure 1. Specific process of the semi-supervised and transfer learning algorithm.

The basic idea of the semi-supervised transfer learning algorithm is to first use the small number of labeled samples in the target domain as training data to train two different classification models, namely the SVM model based on transfer learning and the semi-supervised co-training model; then we select some samples from source domain, use the SVM model based on transfer learning to evaluate the transfer ability of each sample, delete the samples that are not related to the target domain and obtain candidate sample set. After this we select some unlabeled samples from the target domain, use the semi-supervised co-training model to evaluate the confidence of each sample, delete the samples with lower confidence and add the remaining samples to the candidate sample set.

In the process of selection training samples, not only must we consider the transfer ability, but we must assess the confidence of the sample's category. Then we add the samples satisfying the conditions to the training set. The above sample selection method is based on the basic assumptions of transfer learning and semi-supervised learning. By repeating the process, the number of labeled samples in target domain D_t can be continuously increased.

4. Experiments

4.1. Experiment Settings

4.1.1. Experiment Environment

We build the simulation experiment development environment of Windows7 + Matlab2017b + Libsvm3.22, where Libsvm3.22 is used to implement the SVM model as the base classifier. Its kernel function is based on the radial basis function $\exp(-\frac{|x-x_i|^2}{\sigma^2})$. On this basis we use Matlab to realize the transfer learning and semi-supervised learning method in this paper.

4.1.2. Experiment Data

We use the characteristic parameters such as pulse amplitude(PA), carrier frequency (CF), pulse width (PW), pulse repetition interval (PRI) and angle of arrival (AOA) to simulate generating the emitter data of six system-like radars. For the signal parameters, they are set at the same intermediate frequency: 10 MHz, and the sampling frequency is 100 MHz. 1000 signal samples are generated using the above five pulse description words for radar 1, radar 2 and radar 3, respectively, and a total of 3000 signal samples are as known radar emitter data corresponding to the source domain data above.

In addition, 1000 signal samples are generated for radar 4, radar 5 and radar 6, respectively, and a total of 3000 signal samples are as unknown radar emitter data corresponding to the target domain data above. The mean values and standard deviations after normalization of the known radar emitter data and the unknown radar emitter data are significantly different, so they no longer satisfy the assumption of the same distribution, which can be used to verify the transfer learning and semi-supervised learning method. The details of the experiment data are shown in Tables 1 and 2.

Table 1. Known radar emitter data.

Known Radar	PA	CF/MHz	PW/μs	PRI/μs	AOA/°
radar 1	[6, 16]	[2019, 2020]	[1.1, 1.3]	400/500/550	[46, 48]
radar 2	[2, 12]	[2150, 2250]	[0.3, 0.5]	300/350/400	[62, 64]
radar 3	[16, 20]	[3121, 3333]	[7.1, 7.2]	800/830/860	[66, 68]

Table 2. Unknown radar emitter data.

Unknown Radar	PA	CF/MHz	PW/μs	PRI/μs	AOA/°
radar 4	[12, 14]	[2545, 2546]	[0.2, 0.4]	710/730/770	[28, 30]
radar 5	[5, 8]	[2763, 2773]	[0.6, 0.8]	240/280/320	[52, 54]
radar 6	[23, 31]	[2855, 3003]	[4.7, 4.8]	600/640/660	[48, 50]

A radar signal sample is written as $W_i = [PA_i, CF_i, PW_i, PRI_i, AOA_i]^T$. The distribution of the specific parameters of the radar is shown in Figure 2, (a) describes the entire data set from the perspective of parameters PA and AOA, and (b) describes the entire data set from the perspective of parameters of CF, PW and PRI.

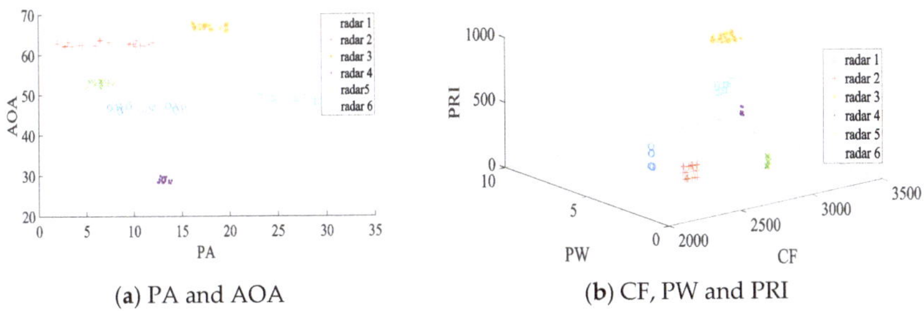

(**a**) PA and AOA (**b**) CF, PW and PRI

Figure 2. Distribution and change characteristics of the five parameters of pulse amplitude (PA), carrier frequency (CF), pulse width (PW), pulse repetition interval (PRI) and angle of arrival (AOA).

4.2. Interaction between Transfer Learning Method and Semi-Supervised Learning Method

The experiment uses the known radar emitter data as labeled samples for auxiliary training, and the unknown radar emitter data as unlabeled samples to be identified. The number of unlabeled samples and labeled samples can be adjusted to verify the interaction between the transfer learning method and the semi-supervised learning method.

First, we keep the number of labeled samples unchanged, and adjust the number of unlabeled samples to verify the impact of the semi-supervised learning method on the transfer learning method. The results are shown in Figure 3. It can be seen from the experiment results that when the number of unlabeled samples is zero, that is, we only carry out transfer learning without semi-supervised learning, the identification accuracy is 11.3% lower than the optimal identification accuracy.

When the number of unlabeled samples is slowly increasing, the identification accuracy will also continue to rise, indicating that the unlabeled samples help to make the transfer learning method select high-similarity samples from the unknown radar emitter data; that is, the semi-supervised learning method is positively correlated with the transfer learning method, and has not weakened it. As the number of unlabeled samples increases further, the identification accuracy will gradually stabilize, indicating that the high-similarity samples in the unknown radar emitter data have been completely screened out, and the optimal recognition rate can reach 93.6%.

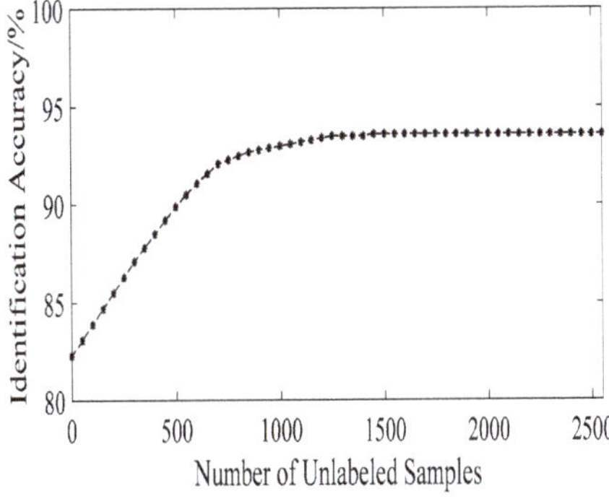

Figure 3. Impact of the semi-supervised learning method on the transfer learning method.

Secondly, we keep the number of unlabeled samples unchanged, and adjust the number of labeled samples to verify the impact of the transfer learning method on the semi-supervised learning method. The results are shown in Figure 4. It can be seen from the experiment results that when the number of labeled samples is zero, that is, we only carry out semi-supervised learning without transfer learning, the difference between the maximum identification accuracy and the minimum identification accuracy in the classification identification results reaches 17.9%, indicating that the use of semi-supervised learning alone makes the model less stable. When the number of labeled samples is slowly increased, the difference between the maximum identification accuracy and the minimum identification accuracy in the classification identification results will continue to drop to 4.3%, indicating that the transfer learning method helps to make self-correction of the semi-supervised learning method.

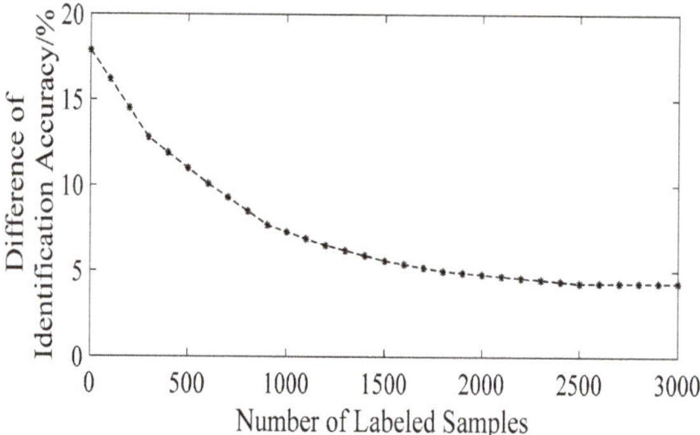

Figure 4. Impact of the transfer learning method on the semi-supervised learning method.

It can be seen from the above experiment results that the semi-supervised and transfer learning method proposed in this paper can comprehensively utilize the information of unlabeled samples and labeled samples. When the number of unlabeled samples is greater than 1000, and the number of labeled samples is greater than 1500, the performance of the model will tend to be stable and achieve the highest identification accuracy. Therefore, in the following we use 1500 known radar emitter samples and 1000 unknown radar emitter samples to train the model for contrast experiments.

4.3. Contrast Experiments

In order to further verify the effectiveness of the proposed method, we train the base classifier SVM model, the SVM model based on transfer learning, the SVM model based on semi-supervised learning and the SVM model based on semi-supervised and transfer learning respectively to identify the unknown radar emitter samples. In addition, in order to verify the adaptability of the model to the measurement error, we introduce an error deviation level test algorithm [21]. The specific experiment results are shown in Figure 5.

From the contrast experiment results, it can be known that when only using the base classifier SVM model for identification, the identification accuracy obtained is less than 55%. The main reason is that the known radar emitter data and the unknown radar emitter data do not satisfy the same-distribution hypothesis, resulting in an inability to obtain a valid classifier. When using the SVM model based on semi-supervised and transfer learning for identification, the optimal identification accuracy can be achieved within a certain measurement error range. Identification accuracy can reach more than 90% in the measurement error range of 15%, indicating that the method has good noise adaptability, and is obviously superior to the SVM model based on transfer learning, and the SVM model based on semi-supervised learning. The main reason is that the semi-supervised and transfer learning can make full use of sample information to achieve good performance without a lot of iteration.

The identification accuracy obtained by the SVM model based on transfer learning is slightly better than that obtained by the SVM model based on semi-supervised learning. The main reason is that there are not many available training samples in the target domain, which leads to the fact that only using the semi-supervised learning method cannot enhance the training effect. When the measurement error is greater than 10%, the identification accuracy of the transfer learning method and the semi-supervised learning method will be significantly reduced, indicating that their noise adaptability is not good.

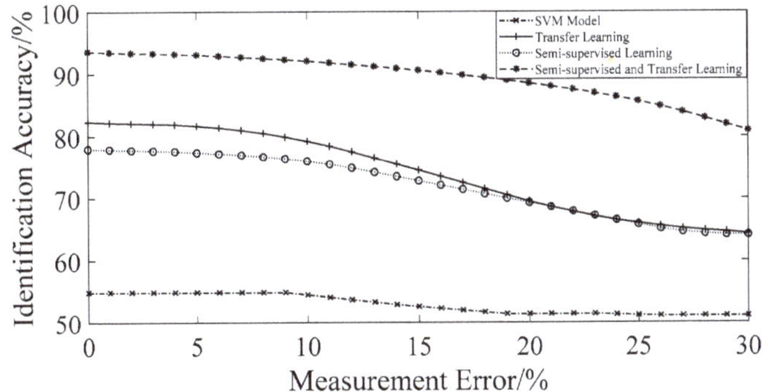

Figure 5. Contrast experiment results.

4.4. Results Discussion

For the radar emitter identification task, deep learning models can often achieve the best results. Therefore, in this section, we construct the CNN model [6] and the U-CNN model [7] to compare with our method proposed in this paper. In the two deep learning models, radar pulse description words are used to represent radar signals, and as input to the model, which is the same as the processing of our method, so it is appropriate to compare CNN, U-CNN and our method together. The specific experiment results of different models are shown in Table 3. In the traditional identification scenario, that is, where we only use the labeled samples in source domain to train the models and then test on the source domain data, U-CNN can achieve the best performance. Its identification accuracy is up to 98.5%, while the identification accuracy of our method is only 95.3%. In the unknown identification scenario, that is, wherein we use the labeled samples in source domain and the unlabeled samples in target domain to train the models and then test on the unknown radar emitters in target domain, the identification accuracy of CNN and U-CNN decrease sharply. However, our method can still reach an identification accuracy of 91.6%. The experiment results show that compared with the currently most popular deep learning models, although our method still has disadvantages in the traditional identification scenario, it can achieve the best performance when facing unknown radar emitters.

Table 3. Identification accuracy of different models.

Model	Identification Accuracy	
	Traditional Scenario	Unknown Scenario
CNN	98.1%	72.2%
U-CNN	98.5%	76.3%
our method	95.3%	91.6%

5. Conclusions

In the radar emitter identification task, the traditional methods are often difficult to identify unknown radar emitters. Aiming at the problem, this paper proposes an unknown radar emitter identification method based on semi-supervised and transfer learning. The transfer learning method can solve the problem that the training data and testing data do not satisfy the same-distribution hypothesis, and the semi-supervised learning method can utilize the information of unlabeled samples to enhance the final training effect. Simulation experiments show that the proposed method can achieve an identification accuracy of 91.6% in the measurement error range of 15%, which is 15.3% higher than the deep learning model in the unknown identification scenario. The next step is to continue to optimize the model and lighten it for automatic compression, in order to minimize the running time of our method.

Author Contributions: Formal analysis, N.T.; Funding acquisition, R.F.; Investigation, X.C.; Supervision, G.W.; Visualization, Z.L.; Writing—review & editing, Y.F.

Funding: This research received no external funding.

Conflicts of Interest: The authors declare no conflict of interest.

References

1. Wiley, R.G. *ELINT: The Interception and Analysis of Radar Signals*, 1st ed.; Artech House: London, UK, 2006; p. 451.
2. Jin, L.; Zhang, X.; Gong, J.Z.; Tang, J.T.; Ren, Z.Y.; Li, G.; Deng, Y.L.; Cai, J. Signal-Noise Identification of Magnetotelluric Signals Using Fractal-Entropy and Clustering Algorithm for Targeted DE-Noising. *Fractals* **2018**, *26*, 1840011.
3. Li, J.C.; Ying, Y.L. Radar signal recognition algorithm based on entropy theory. In Proceedings of the 2014 2nd International Conference on Systems and Informatics (ICSAI), Shanghai, China, 15–17 November 2014.
4. Yang, Z.T.; Qiu, W.; Sun, H.J.; Nallanathan, A. Robust Radar Emitter Recognition Based on the Three-Dimensional Distribution Feature and Transfer Learning. *Sensors* **2016**, *16*, 289. [CrossRef] [PubMed]
5. Zhou, Z.W.; Huang, G.M.; Chen, H.Y.; Gao, J. Automatic Radar Waveform Recognition Based on Deep Convolutional Denoising Auto-Encoders. *Circuits Syst. Signal Process* **2018**, *37*, 4034–4048. [CrossRef]
6. Cain, L.; Clark, J.; Pauls, E.; Ausdenmoore, B.; Clouse, R.; Josue, T. Convolutional neural networks for radar emitter classification. In Proceedings of the 2018 IEEE 8th Annual Computing and Communication Workshop and Conference (CCWC), Las Vegas, NV, USA, 8–10 January 2018; pp. 79–83.
7. Sun, J.; Xu, G.L.; Ren, W.J.; Yan, Z.Y. Radar emitter classification based on unidimensional convolutional neural network. *IET Radar Sonar Navig* **2018**, *12*, 862–867. [CrossRef]
8. Kong, M.X.; Zhang, J.; Liu, W.F.; Zhang, G.L. Radar Emitter Identification Based on Deep Convolutional Neural Network. In Proceedings of the 2018 International Conference on Control, Automation and Information Sciences (ICCAIS), Hangzhou, China, 24–27 October 2018; pp. 309–314.
9. Wang, X.B.; Huang, G.M.; Zhou, Z.W.; Tian, W.; Yao, J.L.; Gao, J. Radar Emitter Recognition Based on the Energy Cumulant of Short Time Fourier Transform and Reinforced Deep Belief Network. *Sensors* **2018**, *18*, 3103. [CrossRef] [PubMed]
10. Shi, Z.H.; Hao, H.; Zhao, M.H.; Feng, Y.N.; He, L.F.; Wang, Y.H.; Suzuki, K. A deep CNN based transfer learning method for false positive reduction. *Multimed. Tools Appl.* **2018**, *78*, 1–17. [CrossRef]
11. Deng, H.; Ma, C.; Shen, L.J.; Yang, C.W. Semi-Supervised Learning Using Autodidactic Interpolation on Sparse Representation-Based Multiple One-Dimensional Embedding. *Int. J. Wavelets Multiresolution Inf. Process.* **2019**, *17*, 1950013. [CrossRef]
12. Dainotti, A.; Pescape, A.; Claffy, K.C. Issues and future directions in traffic classification. *IEEE Netw.* **2012**, *26*, 35–40. [CrossRef]
13. Zhu, Z.F.; Zhu, X.Q.; Guo, Y.F.; Xue, X.Y. Transfer incremental learning for pattern classification. In Proceedings of the 19th ACM Conference on Information and Knowledge Management (CIKM), Toronto, ON, Canada, 26–30 October 2010.
14. He, Q.; Li, B.; Shen, B.; Xia, Y. Cross-Project Software Defect Prediction Using Feature-Based Transfer Learning. In Proceedings of the 7th Asia-Pacific Symposium, Wuhan, China, 6 November 2015.
15. Montalto, A.; Marinazzo, D.; Kugiumtzis, D.; Nollo, J.; Faes, L. Comparing Model-Free and Model-Based Transfer Entropy Estimators in Cardiovascular Variability. In Proceedings of the Computing in Cardiology 2013, Zaragoza, Spain, 22–25 September 2013.

16. Jefferson, T. Boosting Degree Completion and Transfer Rates: An Examination of Counseling/Advising Using the Relationship-Based Model. *Online Submiss* **2010**, 36. Available online: https://eric.ed.gov/?id=ED538968 (accessed on 11 December 2019).
17. Xu, X.P.; Zhu, X.W.; Liu, Q.Q. Can Self-Training in Mindfulness-Based Cognitive Therapy Alleviate Mild Depression Among Chinese Adolescents? *Soc. Behav. Personal. Int. J.* **2019**, *47*, 4. [CrossRef]
18. Karamanolakis, G.; Hsu, D.; Gravano, L. Leveraging Just a Few Keywords for Fine-Grained Aspect Detection Through Weakly Supervised Co-Training. *arXiv* **2019**, arXiv:1909.00415.
19. Goodfellow, I.J.; Pouget-Abadie, J.; Mirza, M.; Xu, B.; Warde-Farley, D.; Ozair, S.; Courville, A.; Bengio, Y. Generative adversarial nets. In Proceedings of the Conference and Workshop on Neural Information Processing Systems, Montreal, QC, Canada, 8–13 December 2014; pp. 2672–2680.
20. Kipf, T.N.; Welling, M. Semi-Supervised Classification with Graph Convolutional Networks. *arXiv* **2016**, arXiv:1609.02907.
21. Shieh, C.S.; Lin, C.T. A vector neural network for emitter identification. *IEEE Trans. Antennas Propag.* **2002**, *50*, 1120–1127.

Article

Design Limitations, Errors and Hazards in Creating Decision Support Platforms with Large- and Very Large-Scale Data and Program Cores

Elias Koukoutsis [1,*], Constantin Papaodysseus [1], George Tsavdaridis [2], Nikolaos V. Karadimas [3], Athanasios Ballis [4], Eirini Mamatsi [1] and Athanasios Rafail Mamatsis [1]

1 School of Electrical and Computer Engineering, National Technical University of Athens, 15708 Athens, Greece; CPapaod@cs.ntua.gr (C.P.); eirinimamatsi@gmail.com (E.M.); trmamatsis@mail.ntua.gr (A.R.M.)
2 Hellenic Army General Staff, 15708 Athens, Greece; GTsav@central.ntua.gr
3 Division of Mathematics and Engineering Science, Department of Military Science, Hellenic Army Academy, 21100 Nafplion, Greece; nkaradimas@sse.gr
4 Department of Transportation Planning and Engineering, National Technical University of Athens, 15708 Athens, Greece; abal@central.ntua.gr
* Correspondence: E.Koukoutsis@ece.ntua.gr

Received: 26 October 2020; Accepted: 10 December 2020; Published: 14 December 2020

Abstract: Recently, very large-scale decision support systems (DSSs) have been developed, which tackle very complex problems, associated with very extensive and polymorphic information, which probably is geographically highly dispersed. The management, updating, modification and upgrading of the data and program core of such an information system is, as a rule, a very difficult task, which encompasses many hazards and risks. The purpose of the present work was (a) to list the more significant of these hazards and risks and (b) to introduce a new general methodology for designing decision support (DS) systems that are robust and circumvent these risks. The core of this new approach was the introduction of a meta-database, called teleological, on the base of which management, updating, modification, reduction, growth and upgrading of the system may be safely and efficiently achieved. The very same teleological meta-database can be used for the construction of a sound decision support system, incorporating elements of a previous one at a future stage.

Keywords: very large-scale decision support systems; very large-scale data and program cores of information systems; meta-database; teleological meta-database; thematic list; indicators list; computational methods list; geographically dispersed systems; external sources

---

1. Introduction

Nowadays, there is a considerable number of public and private operators, who make extensive use of decision support systems (DSSs or DS systems) in a surprising large number of operational procedures and businesses. These operations include:

- Vehicle or vessel fleet management [1–3];
- Marine services and policies (testing, inspection and certification services in the areas of quality, health and safety, as well as security, environmental considerations and relevant policy impacts, etc.) [4–6];
- Management of goods in road, rail, sea and intermodal transportation. These operations are very complex, involving public and/or private train, truck and ship owner companies, goods owners and buyers, transport operators, intermodal terminal operators, infrastructure providers, safety and security operators, etc. [7–9];

- Handling of emergencies and crisis [10–13];
- Complex engineering tasks and, in general, large and complex sequences of operations and/or tasks covering very extended geographical areas (e.g., aircraft and sea vessel design and very large and/or complicated production lines in factories and many more) [14–17];
- Transport planning (estimation of important parameters than cannot be measured, forecasts, policy making, etc.) [18–21];
- Cost-benefit analyses [22–24];
- Healthcare, a domain that is fundamental for human society, especially nowadays (in 2020) due to the COVID-19 pandemic [25–27].

These examples of operations constitute only a small number of those encountered in practice. In addition, in most cases, each associated DSS employs sets of very complex computational models and algorithms. In essence, a DSS is a special kind of a very complex information system (IS), where an information system is an integrated set of components for collecting, storing and processing data and for providing information, knowledge and digital products [28].

The present manuscript is organized as follows:

In Section 2, the authors deal with the recent tendency/necessity to build very large-scale decision support systems. In fact, it is argued that the advances in information and communication technology have given the opportunity to engineers to try to develop DSSs that tackle very big and/or particularly complicated practical problems.

In Section 3, a number of serious difficulties and problems are stated, in connection to specific properties/characteristics of a very large-scale decision support system. These properties concern the following:

- The ability to efficiently update a DS system;
- The flexible expansion of the DSS;
- The consistent reducibility of an information system;
- The ability to make changes to the information system;
- The upgrading of the DSS.

In Section 4, crucial characteristics of contemporary, mainstream methodologies for creating large-scale decision support systems are reported. It is argued that any ad hoc development of a DS system is bound to suffer from serious intrinsic problems, which only a consistent fully methodological approach can resolve.

In Section 5, the authors introduce a methodology that circumvents all the aforementioned problems.

In Section 6, the authors state the reasons for which the introduced methodology resolves these problems associated with very large-scale DSS development.

Finally, in Section 7, the conclusions of the present work are stated.

2. The Tendency to Increase the Volume and Complexity of Data and Program Collection of the DSS

A considerable number of contemporary DSS applications eventually ask for very large data and program collections. The tendency to increase the size and complexity of these collections is due to a variety of factors, such as:

(a) Advances in information technology and communications (ITC) have created a market for DSSs with very large-scale data and program collections.

In fact, progress in software design and implementation makes possible the almost instant handling of huge data sets. For example, the conceptual development of balanced, N-ary trees allows for accessing a specific entry among trillions of entries in four or five very fast steps only. In addition, contemporary hardware allows for creating and managing/handling "storages", that is, devices that may contain thousands of disks (hard disks or DSSs).

(b) The same ITC advances have opened the way to engineers to attempt to develop DSSs that tackle very big and/or complicated practical problems. Two related examples follow:

- The effort for developing the European Transport Information System for Policy Support (ETIS).

Maintaining an efficient transport infrastructure and formulating a common transport policy are critical elements for the economic and social development of the European Union (EU). In 1994, the EU Commission adopted a comprehensive proposal for the "Trans-European Networks" (TEN-T) guidelines [29], which are plans for improving the performance and further development of the complex EU infrastructure. The first version of these guidelines covered 70,000 km of rail tracks (including 22,000 km of new and upgraded tracks for high-speed trains), 58,000 km of roads (including 15,000 km of new roads), corridors and terminals for combined transport, 267 airports of common interest, networks of inland waterways and many seaports. It is evident that these numbers have greatly increased since the adoption of TEN-T guidelines.

The effort for performance improvement of the Trans-European Networks and the application of a successful transport policy requires significant information concerning the current status of the networks and the corresponding inhabited regions. More specifically, this required information includes passenger and freight transport volumes, traffic congestion, environmental impacts of transport, etc. It is also necessary to consider reliable forecasts for the time evolution of the aforementioned factors. The relevant information exists in a large number (more than 500) of dispersed, heterogeneous and autonomous databases throughout the entire Europe. In addition, more relevant information is acquired from the results of an ensemble of pertinent scientific studies. The European Transport Information System for Policy Support has been envisaged as a tool for the support of decision making on transport policies and policies related to the Trans-European Transport Networks. ETIS must accommodate policy-related information in a repository that has to be kept up to date by experts. In ETIS, policy questions are linked to the relevant data sets of the sources via a hierarchical structure of policy criteria, policy indicators and data variables [20].

- A second example is the information system for the port of Rotterdam. This system includes a huge amount of geographical information. It is noteworthy that the port of Rotterdam incorporates more than 5000 docs, an airport, a considerably large storage area for alumina and abundant apron spaces for temporary storage of goods and containers.

The IS of the port of Rotterdam, called Port Community System (PCS), provides services that focus on all port sectors, such as containers, break bulk, dry bulk and liquid bulk. Anyone in the logistics chain can easily and efficiently exchange information through these services. Moreover, the PCS offers a package of properly tailored services to each of the target groups of clients and operators.

From the entire previous discussion, it is evident that the information involved in the PCS is huge and very complex [30].

(c) The inclusion and use of data, which are directly or indirectly georeferenced (map-related data), in a DSS. In the first case, the direct one, the exact co-ordinates of an object are explicitly provided. On the contrary, in the second case, the indirect one, there is a link to an object, which determines the data co-ordinates. For example, if an accident takes place at a specific point of a certain road and then, if the exact co-ordinates of the accident location are registered in an IS, this event is directly georeferenced. On the other hand, if the registered information is something like the following: "the accident happened at the 23rd kilometer of the road from city A to city B", then the event is indirectly georeferenced.

Usually, the main bulk of the information is indirectly georeferenced. A surprisingly large part of the information that should be included in such a DSS is georeferenced, even though this georeferencing is well hidden at a first glance. For example, in most nations, such as the USA, the laws and directives may drastically change when crossing state boarders. Even in Greece, where the legislation is much more uniform, when one travels from an island to the mainland, one faces various changes in authority jurisdiction and legislation in the interior of the island, the mainland, at ports and on the sea (near-port area, national waters and international waters).

It is evident that the requirement that a large-scale decision support system handles and uses map-related information may greatly increase the volume of its data and program core.

(d) Finally, special DSS architecture requirements, like the following:

- Distributed and/or geographically dispersed systems, where properly selected duplication of part of the information and programs must be implemented, in order to increase the efficiency of the local systems. For example, a very frequently accessed part of information by all sub-systems must be preferably kept in every local sub-system.
- Umbrella-type systems that receive and process information in real time from a large number of heterogeneous, autonomous and, possibly, legacy systems. Sometimes, these systems may cover a large part or the whole of a continent. Such is the case of a European system for maritime surveillance, the complexity and size of which is evident.

3. Severe Difficulties Appearing in the Actual Deployment of a Very Large-Scale Decision Support System

We use the general term "large data collection" in order to describe the extended data core of a large-scale decision support system. The data collection may include dozens or even hundreds of groups of loosely related data sets or databases, the content of which may be thematically divergent.

In this work, we will often use examples of large-scale decision support systems for policy support and decision making in the area of transport. We will do this for two main reasons:

- In many transport DS systems, practically all problems referred to in the present work have already emerged, due to the huge size and complexity of the involved data. As a first example, we state the large number (more than 50) of Federated States in the USA. An analogous second example is the European Transport Information System, which is a DSS for supporting transport policy decisions in all EU countries.
- Engineers and other scientists that have developed the pioneering versions of the transport DS systems have faced a considerable number of serious difficulties and problems, many of which still remain unsolved. It must be emphasized that for a considerable number of these difficulties and problems, the severity of the resulting hazards may be anticipated.

We would like to mention the following, widely used terminologies in transport DS systems: The data and program collections are often called "repositories", "observatories" or "transport observatories". Frequently, when the transport observatories cover exclusively the needs of a single country, they are called "national models" [31–33].

3.1. Early Design Errors and Serious Difficulties Due to the Large Data and Program Collection Characteristics

In this Subsection, we present some serious difficulties/problems, inherent to large data and program collection characteristics. We will describe below some of the most severe difficulties of this type:

(a) A prevailing design mentality in connection with the first transport DS systems was the following: "Get all related or even loosely related data first and then organize them". Practice, though, proved that this design mentality leads to disasters. In fact, it turned out that:

- This way of data acquisition does not guarantee a precise/consistent, methodological collection of data. On the contrary, the random, heuristic and ad hoc data collection, as a rule,

results in a core with data and programs that are non-interoperable, internally inconsistent, practically impossible to organize or update properly and impossible to upgrade.
- Due to the aforementioned problems concerning the selected data, the core of the corresponding DSS practically collapses under the weight of the non-organizable data collections.

Consequently, this design methodology has been rather quickly abandoned.

(b) It must be pointed out that, after this early approach, the designers of a large-scale decision support system apply carefully designed data acquisition methodologies. However, it turns out that in practical cases, there are several characteristics, inherent to the elements of large data collections, which create severe difficulties and problems to the designers of the information system to be used as a platform of the DSS. Some of these characteristics will be presented immediately below.

(i) As a rule, in a very large DS system, the data are multi-thematic; hence, it is logical to expect a great variability of definitions, meanings and contents associated with the data.
(ii) The aforementioned variability is frequently accompanied by a great diversity of forms and formats of the elements of the data collection. This diversity practically always asks for a different handling of each subset of similar data.
(iii) (Complexity is another serious problem associated with large-scale data collections and the corresponding DS systems. The term "complexity" is used in order to describe data with a particularly great number of interconnections, inter-relations and interdependencies.
(iv) Data polymorphism can easily destroy an arbitrary DS system. For example, the most common computer object in transport, the "transport link", may literally have tenths or even hundreds of different definitions, meanings and data contents. Thus, the link between two cities A and B may refer to:

- Various road connections, where each road has its own characteristics (e.g., a highway with a given number of lanes with or without tolls, a national road, a secondary street, etc.);
- Various train connections, each one frequently with its own characteristics, such as high-velocity trains (TGV), intercity trains, local trains, commercial trains, etc.;
- Airplane connections, with airports having different connections with the city center;
- Seaways, where each way frequently includes ferries, container carriers, bulk carriers, tankers, cruisers, etc.;
- Inland navigation, etc.

In addition, the information concerning a single object in the DSS drastically depends on the purpose of the use of this object. This fact makes large-scale decision support system designers employ ad hoc solutions, which usually make upgrades and thematical shifts much more difficult or even impossible. An efficient solution to this problem will be presented later in Sections 5 and 6 of the present work.

(c) In many cases, large data and program collections must be distributed in several systems, which may be geographically dispersed. Achieving an efficient operation of each such sub-system, as well as maintaining a proper communication among the individual systems, while keeping the large data and program core healthy and consistent, is a difficult task indeed.
(d) The huge numbers of external data sources of large DS systems: In a considerable number of cases, the large DSS data core must handle data from a huge number of different, heterogeneous and autonomous external sources, such as databases and/or various data collections. For example, the European Transport Information System for Policy Support (ETIS) receives data from more

than 500 external sources [20], which are indeed heterogeneous and autonomous. Handling this number of different sources is a very difficult task and, in most cases, tackling this task is by no means automatic. Indeed, usually, a large initial effort by experts is required, in order to create a first version of an interoperable, internally consistent data and program core [34].

(e) In a large data collection, a considerable number of differences in the meaning and/or definition of common variables, coming from different sources, may appear. Thus, the task of defining the exact meaning and form of the variables that are common to many sources of the DS system may prove to be extremely difficult.

For example, in the first stages of the ETIS design, it has been proved that the difficulty of defining the term/variable "long distance trips" [35] was almost insurmountable. In fact, for the Scandinavians, this term referred to a distance greater than 100 km, while such an option practically excluded trips in the Netherlands entirely. In addition, other countries expressed a strong desire to associate with this term the age of the passengers and/or to incorporate in it whether the relevant trip is transit or not. Today, one may think that the term "long distance trips" must also include as a sub-variable the information whether the vehicle (automobile, motorcycle or train) is electric or conventional.

Thus, it not a surprise that a new, complete definition of the term "long distance trips" was given by the Transport General Directory and that Eurostat issued a directive to all EU Members to conform with this new definition and to always provide all relevant information.

(f) The next difficulty concerns the "level of detail" (abbreviated as LoD) or "granularity". We shall try to clarify the content and importance of the term "level of detail" by means of the following example, concerning Google Maps. In fact, this application starts from offering a global earth projection. Then, the user may gradually increase the level of detail at an arbitrary point of the projection, according to his/her desire. At each such selection of the user, a more detailed map of the associated area appears on the screen. At certain levels, new geographical objects appear on the screen. For example, one may select a specific country and see its map. Subsequently, one may choose an area of this country and see the corresponding map in more detail. After a sequence of successive selections, where each one of them offers a greater level of detail, one may suddenly see some cities as points and some sketches of roads; hence, new objects appear. Further increase of the detail may offer more and more dense maps of the desired area. If one selects the option "satellite", more and more detailed images of the selected area appear on the screen. In the higher detail level, the user may see individual public buildings and constructions, houses and details of them.

In the first steps of large DS system implementation, many designers considered that it is sufficient to include the higher level of detail only in the information system. However, the two aforementioned examples, but also many other DS systems that do not necessarily include geographical information, demonstrate that the approach "include the highest level of detail only" is completely erroneous. On the contrary, as it will be analytically presented in Sections 5 and 6 of the present work, a sound and systematic method for handling the levels of detail in a large DS system is to include all necessary information, associated with any level of detail, according to each issue and problem that the DSS has to tackle. In other words, the levels of detail in each application of the DSS strongly depend on the issue and problem in hand; therefore, the corresponding information system must be designed by keeping this issue dependence in mind.

3.2. Extreme Problems in Deploying, Managing, Maintaining, Updating and Upgrading Very Large-Scale Decision Support Systems

The deployment of a (very) large-scale decision support system meets with severe difficulties, due to the reasons described immediately below, which are in strong accordance with the content and the observations of [36,37]:

(a) Since the DSS platform is not sustainable in many cases [38], an effort for redesigning the DSS must be repeated after a short time period, when considerable changes must be accommodated. We would like to emphasize that the time period, after which it is necessary to redesign a certain DS system [39] (p. 4) [40,41], may frequently be as short as few years only, for example, 4 or 5 years.

(b) Serious difficulties in updating a large-scale decision support system: In such a system, the great number of the involved data, the significant complexity and polymorphism of them, the heterogeneity and dispersion of the data sources, as well as the fact that the various source data bases have been designed with a different mentality/approach, make the necessary, regular updating of the overall information system too difficult or even impossible. In fact, the problem is far more severe. The aforementioned factors render DSS upgrading much more difficult; as a rule, upgrading comprises improvement of existing data sets and programs, as well as incorporation of new data sets and programs, so that additional problems can be tackled by the DSS.

(c) Upgrading a large-scale decision support system is, very frequently, impossible: We would like to point out that, when a serious upgrading of a system is required, then, as a rule, a new DS system is frequently designed ad hoc, without a systematic methodology; this approach asks for a migration of the data and program core of the old system to the new one, a task which is extremely difficult and, in certain cases, impossible to achieve. Evidently, the new DS system will definitely manifest the very same problems as the old one, concerning updating and upgrading.

(d) The DSS lifecycle may strongly depend on the duration of the incumbency of the decision authorities: It is very well known that, in western-type democracies, the authorities who take decisions frequently change in a smooth way. This fact may render the time duration of the incumbency of the policy makers comparable to the time period, which is necessary for the deployment of a new or improved DS system. Thus, for example, very often, the policy priorities may drastically change, as a result of changes in the decision authorities. In this way, a DS system that has just begun to work may become partially or totally obsolete. Consequently, the DS system must be drastically changed or even redesigned from the beginning. Hence, it is imperative to design information systems in such a way so that maximum re-use of the existing data and programs can be achieved.

4. Contemporary, Mainstream Methodologies for Creating Large-Scale Decision Support Systems

First of all, a complete/thorough analysis and study, concerning the set of the required indicators, usually precedes the system design. This part of the DSS project is absolutely necessary; otherwise, corrections and additions at a later stage may be difficult or even impossible. The initial study must clearly and comprehensively offer information about the following, which will also prove very useful in the analysis of Sections 5 and 6:

(a) A comprehensive list of the general issues/problems, for which the DS system must provide support. We will use for it the term "thematic list".

(b) A decomposition of each one of the aforementioned issues/problems to subcategories, further subcategories, etc., up to the fundamental, non-separable issues/problems. We will employ for this process the term "thematic decomposition". For example, the general transport issue of "social/environmental impacts of transport" must initially discriminate between the subcategories "harmful impacts" and "beneficial impacts". Then, the subcategory of "harmful impacts" must, among others, include:

    - "Noise";
    - "Congestion";
    - "Pollutant emissions".

For example, a complete associated path could be the following: transport policy issues → environmental impacts of transport → harmful environmental impacts of transport → noise → noise produced at the transport network → elementary/non-separable issue: noise produced by road traffic outside of the urban areas → indicator(s): noise measurements for all links and nodes of the road network outside of the urban areas.

On the other hand, the subcategory "beneficial impacts" must, among others, include:

- "Improved connectivity" (e.g., better island connectivity);
- "Transport time reduction";
- "Growth" (e.g., residential or commercial relocation for a better environment, improvement of trade conditions, economic growth, etc.).

It must be emphasized that the form of an indicator can greatly vary, from a simple set of numbers (e.g., the former Kyoto indicators for environmental pollution) [42–45] to a complete database (e.g., a map indicating the road congestion of an area, with seasonal and daily indications).

(c) An ensemble of methods for either acquiring the indicators or computing them from the relevant data, using proper mathematical models.
(d) A complete list of the data, from which indicators must be evaluated.
(e) A complete list of the corresponding software, that is, programs that will implement the aforementioned abstract objects.
(f) A sufficient/adequate ensemble of "metadata", which are necessary for the description of all indicators, data, acquisition or evaluation methods and the justification of their choice. It must be emphasized that this step is quite often partially considered or even neglected in the design of impressively numerous very large-scale information systems. Designing a very large-scale information system and including a poor/insufficient set of metadata is a grave error that will practically render the system non-maintainable, non-upgradable and non-modifiable.

The related information is, as a rule, a result of tedious efforts of experts, usually belonging to numerous scientific disciplines. Evidently, the work of the experts is supported by information technology and communication (ITS) engineers.

As a next step, ITS experts, using the aforementioned abstract results, implement the following:

- A user interface architectural tier, that is, an adequate number of computer systems to handle the communication of the users with the DS system. Nowadays, this is achieved through the Internet and/or an Intranet, if confidentiality reasons dictate so.
- One or more databases and/or data warehouses specifically designed to handle large ensembles of data. The entire set of these tools is frequently called "data tier". We note that the data tier may include specialized hardware, such as hardware "storages" or other similar components.
- A cluster of computer systems, sometimes called "application tier", implementing the computational methods for the evaluation of the indicators, using data from the previous data tier. Additionally, the ITS engineers may include custom software for satisfying particular needs of the users of a DS system for data visualization and/or specialized processing.
- A set of computer systems for the communication and mediation of the DSS with external data bases.

Contemporary solutions may include even more advanced software tools, such as "web services" and "containers" [46]. Moreover, in recent designs, several servers remain totally "in-memory", in order to achieve highly increased performance; however, this solution is particularly expensive.

Next, suppose that a team of engineers implements a DS system ad hoc, without incorporating a rich set of metadata in it. In fact, without the complete knowledge of the appropriate metadata, the programs and data of an information system are practically entangled black boxes, to which corrections, maintenance, improvements, expansion and updating are difficult, if not impossible to be realized. In addition, a considerable number of users themselves are experts and/or they are supported by teams of experts; consequently, they will definitely need to know the details of acquisition and evaluation of the indicators. Thus, we would like to emphasize that the limiting factor in the design of large-scale decision support systems is not the hardware or software capabilities, but the very organization of the huge information itself.

5. A Novel, Systematic Methodology for Designing, Maintaining and Upgrading a Very Large-Scale Decision Support System

It has been previously stated that the design and implementation of a very-large scale decision support system is realized by means of the following two general stages:

Stage 1: A group of experts, after an extensive analysis of the problems/requirements in hand, ends up with an abstract structure, which determines all due operations of the DS system.

Stage 2: Based on this abstract structure, a group of ICT engineers produces an ad hoc implementation of the sought-for information system. However, as it has been more extensively described in Section 4 of the present work, any ad hoc implementation intrinsically suffers from serious flaws and problems of maintenance and upgrading of the developed system.

Hence, in the present section, a novel, systematic approach that circumvents these difficulties is described. This new approach includes the following steps:

(a) Development of a specialized meta-database, able to store the entire aforementioned abstract structure produced by the experts, using a corresponding template, which is more analytically described below. For this meta-database, we shall use the name "teleological meta-database" [47] or simply "meta-database", for reasons that will become clear in the following steps.
(b) Insertion in this template of the thematic entities/entries of the system and the associated "thematic decomposition", which ends up to the "non-separable objects" of the DSS in hand; these actions and terms will be more extensively analyzed in Section 5.1.
(c) Further filling of the template of the teleological meta-database with all the necessary indicator descriptions. At the same time, the analytic description of the exact method(s) of acquiring or evaluating these indicators must be included in this meta-database.
(d) Further inclusion in the teleological meta-database of the complete list of characteristics of the DSS external sources, together with all necessary information for accessing them.
(e) Incorporation of all the documentation metadata in the teleological meta-database; this documentation must fully cover the entire set of data, programs and other components and actions of the DSS.

We would like to emphasize that we render this teleological meta-database the "heart" of the entire DS system and its operations.

5.1. The Part of the Template that Includes the Compete Thematic Decomposition

As it has been already pointed out in Section 4, the first part of the template concerns the complete thematic list of the topics or issues for which the DSS must provide information/answers. An exhaustive thematic decomposition is applied to this list, expressed via a tree-like graph structure, down to further-non-separable objects/entities. We noticed that in the case of polymorphic data, the final leaves of the thematic decomposition usually highly depend on the sub-problem in hand. In other words, a specific object must be further decomposed, when the DSS tackles a certain problem, say A, while it may be a non-separable entity, when the system offers information concerning another problem, say B.

An abstract visualization of this tree-like, graph structure that performs such a thematic decomposition is depicted in Figure 1. In fact, concerning Figure 1, we would like to point out the following:

- The entire thematic list of the DSS incorporates issues $I_1, I_2, \ldots, I_N$.

- Each such issue is decomposed into successive sub-issues, for example, issue I_1 is at the first level decomposed into sub-issues $SI_{1,1}$, $SI_{1,2}$, … , $SI_{1,K1}$. Similarly, issue I_N is decomposed into sub-issues $SI_{N,1}$, $SI_{N,2}$, … , $SI_{N,KN}$.
- In an analogous manner, sub-issue $SI_{1,1}$ is decomposed into $SI_{1,1,1}$, $SI_{1,1,2}$, $SI_{1,1,3}$, … , $SI_{1,1,M1}$, and so on.
- The decomposition process in connection with each initial thematic issue stops, when all non-separable objects, associated with this issue have been determined. In this way, for each issue I_J (J = 1,2, … , N), a specific number of paths is obtained. We emphasize that the number of these paths as well as the depth of each path, as a rule, are not constant. We employ the symbol NS for the non-separable object of each such path. Thus, for example, issue I_1 is eventually decomposed into F1 non-separable objects, giving rise to "last-leaves" $NS_{1,1}$, $NS_{1,2}$,..., $NS_{1,F1}$.

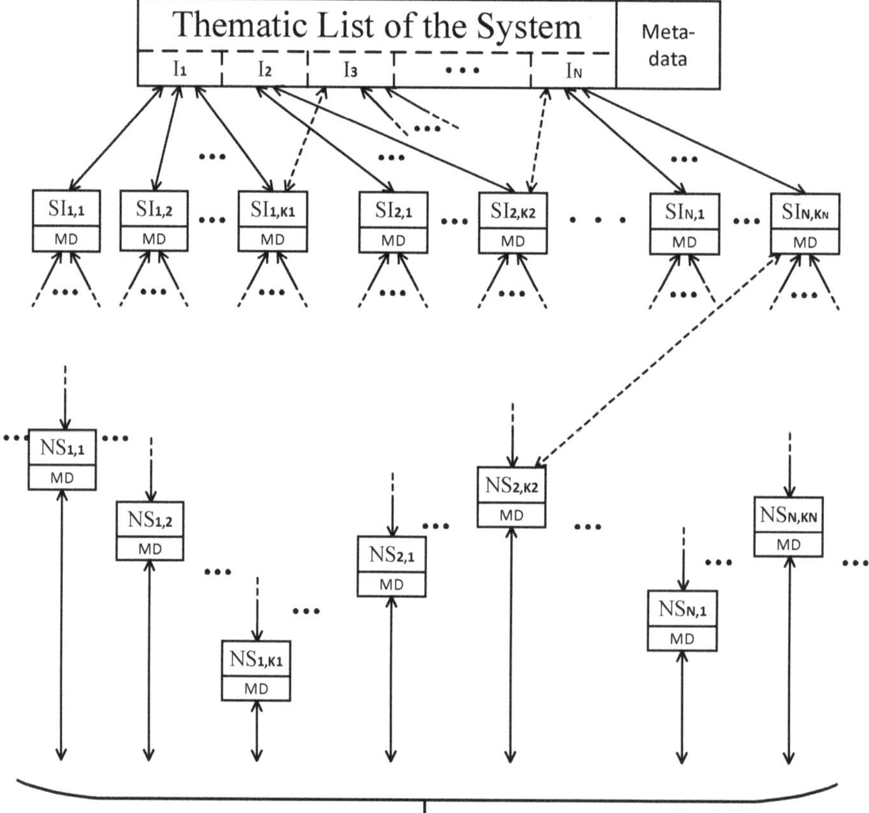

Figure 1. The structure of a thematic decomposition, an analytic explanation of which is presented in Section 5.1, where "MD" stands for "metadata".

We repeat that, often, the length of the path that leads to $NS_{1,1}$ may be different from the length that leads to $NS_{1,2}$, which in turn may be different from the length of the path that leads to $NS_{1,3}$, and so on. These differences in length are indicated in Figure 1 by the fact that the non-separable objects are placed at different levels.

Evidently, issue I_2 ends up to the non-separable objects $NS_{2,1}$, $NS_{2,2}$, … , $NS_{2,F2}$ and so forth. We stress that, in many instances, issue I_J may end up to an arbitrary number of non-separable objects $NS_{J,R}$ following various different paths. In this case, we choose as length of the path that links I_J with $NS_{J,R}$, the length of the path that connects these two entities, which starts at the higher-most level, namely, the one closer to I_J, and so on.

- We have used a double arrow to indicate a connection of two arbitrary objects of the tree-like structure, in order to make clear that each sub-issue must "know" all previous sub-issues and the initial issue(s) that have generated it.
- We would like to emphasize that each issue and sub-issue is naturally linked to a set of metadata.
- The dotted arrows have been used to indicate that the entire structure is not a tree, but a graph, most probably, an acyclic one.
- The gray arrows that originate at the final non-separable objects of this tree-like, graph structure of Figure 1 manifest that this structure is linked to another one, that deals with the indicators and/or the data of the system.

5.2. The Template Part Covering the DSS Indicators and/or Data and Their Acquisition or Evaluation Method(s)

Each issue and/or each sub-issue, described in Section 5.1 and Figure 1, may be linked (and is indeed very frequently linked) to an ensemble of indicators, namely, certain quantitative characteristics of the issues and/or sub-issues in hand. Indicators help the user(s) of the DS system to make decisions as objectively as possible. As we have already pointed out, a simple, but characteristic, example of an important indicator is the noise generated by vehicles at various specific points of a road network [48]. Another crucial indicator is the volume of pollutants emitted by vehicles at these points [49].

We will use the term "the complete list of indicators" to express the entire set of indicators, incorporated in a DS information system.

Each indicator must be associated with (a) the data necessary for its computation and/or (b) the exact method(s) for acquiring or evaluating it.

As a rule, there are three ways of obtaining indicators, namely:

(a) For each indicator (or group of indicators), the DS system uses a specific computing methodology, which employs certain data in order to evaluate the indicator(s).
(b) The indicator (or group of indicators) may come directly from external sources, needing no further (or minimal) processing, and they are stored intact in the DS system.
(c) Finally, there are groups of indicators, which are the results of studies and/or development of external group of experts. These indicator(s) are also stored intact in the DS system.

An abstract representation of all three methods for acquiring or evaluating indicators is shown in Figure 2.

As far as this part of the template is concerned, one more important issue must be emphasized. A very consistent approach to describe a method that computes one or more indicators, so that the DS system remains maintainable and upgradable, is to decompose each such computational method into lesser "functional blocks" and describe all the necessary internal communication of these blocks (i.e., for each block, its input intermediate variables, the origins of these input variables, the output intermediate variables and the destination of these output variables must be described).

This can be better understood my means of a simple, but fictitious, practical example, which is schematically shown in Figure 3. In this example, a certain computational method is decomposed into four (4) well-defined sub-blocks, namely, "model 1", "model 2", "sub-method 1" and "sub-method 2".

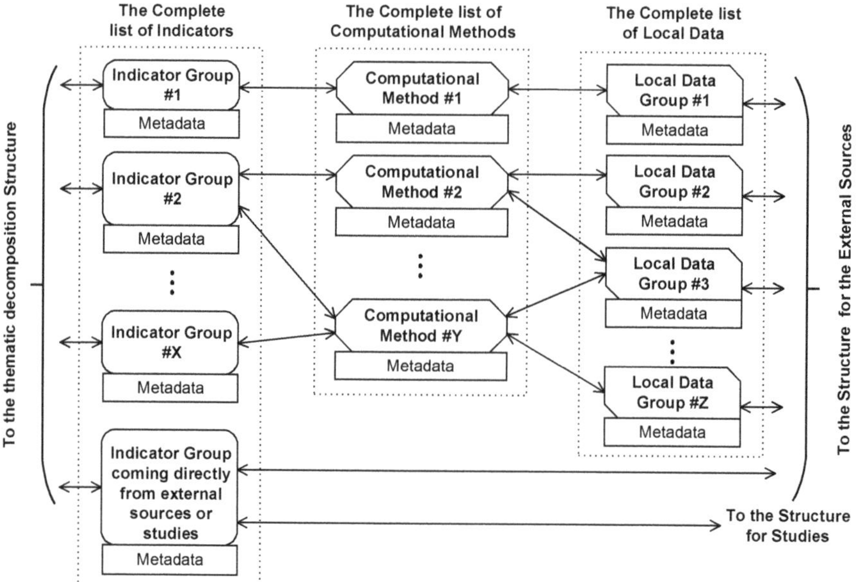

Figure 2. The structure of the meta-database template concerning the indicators, local data and computational methodologies.

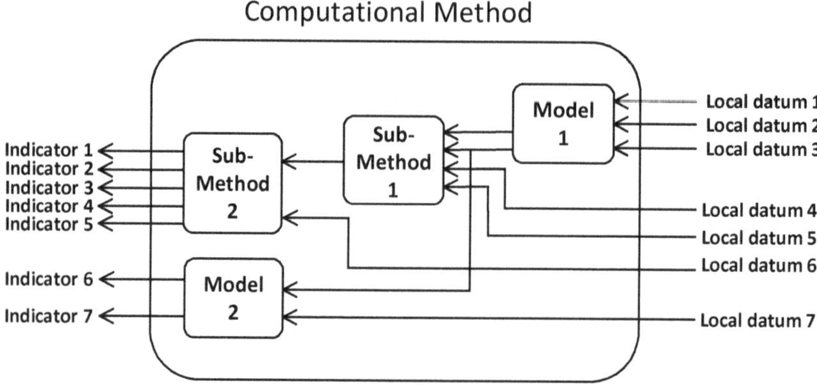

Figure 3. An example of a first step of decomposition of a computational method. Any indicator and/or local datum may be compound, i.e., it may comprise a structure of simpler data.

Then, each sub-method (namely, model 1, ... , sub-method 2) may be further decomposed in an analogous manner and so on. The decomposition stops when the experts that have initially developed the system feel that they have described the entire computational method adequately, so that another expert (or group of experts) can fully understand it at a later stage.

In Figure 4, another example is shown, which is actually encountered in the transport application area. In fact, the example refers to a specific method that computes the noise levels at a number of certain road network traffic links.

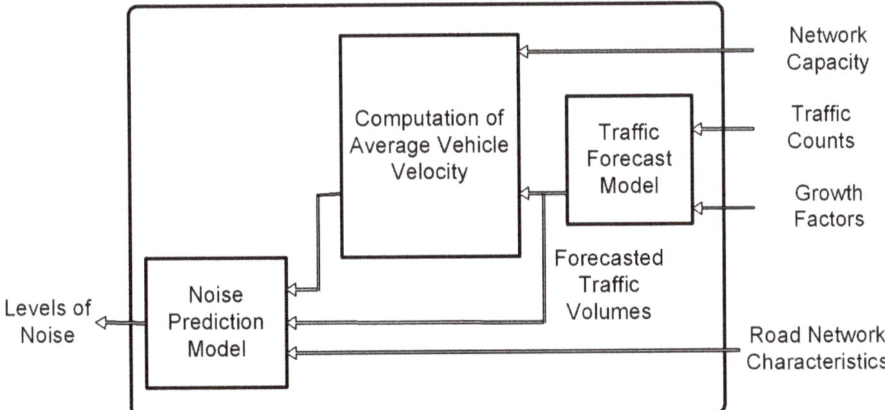

Figure 4. Computation of the levels of noise at certain links of a road network.

First, a traffic forecast model uses traffic counts on the links of the network and traffic growth factors (local data) to forecast traffic volumes for a future planning period. At the next step, a sub-computational method takes these forecasted traffic volumes and road capacity figures and computes the average vehicle velocities. Finally, a noise prediction model takes the computed average vehicle velocities, forecasted traffic volumes at the links of the network, as well as other road network characteristics and estimates the noise levels at these links of the network for the future planning period. The description of the computational methods starts at this level of abstraction. The sub-methods, shown in Figure 4, should be further decomposed. Once more, the decomposition stops when the experts that developed the system feel that the entire description is sufficient.

5.3. Including the External Sources and the Way to Access Them in the Template

We use the term "external sources" to describe the set of geographically dispersed databases linked to the main decision support system; these data bases are, as a rule, autonomous and most probably heterogeneous. Each such database may offer data and/or indicators to the main DS system. In turn, the main system may use these external data or indicators intact or employ them, in order to compute and/or generate other data or indicators. The part of the system that implements the communication of each external source with the main system and, possibly, computes all necessary quantities, is called "mediator"; the corresponding process itself is called "mediation".

As mentioned before, each external source provides a part of the data of the main system and/or a part of the indicators. Consequently, the proposed meta-database must include:

(a) A complete description of the information furnished by the external sources.
(b) A full documentation concerning the method(s), which the mediator employs in order to evaluate the various indicators or data.
(c) The exact methodology and the frequency with which the data and/or indicators reaching the main system must be updated.
(d) Since a part (sometimes considerable) of the indicator ensemble may come from studies/R&D projects, these studies or projects can be considered as an "off-line" kind of external sources. It is necessary to describe the information associated with these indicators adequately, so that future expert users of the DS system can clearly and unambiguously understand the following:

- The quality of the indicators;
- Any possible special characteristics these indicators possess;

- Any limitations of the use of the group of specific indicators.

5.4. Including a Complete, Structured Description of Experts' Knowledge in the Teleological Meta-Database

Figures 1 and 2 depict a fundamental characteristic of the meta-database, expressed by the fact that a "metadata box" is attached to every entity in the template. Each such box should incorporate the relevant knowledge of the experts who built the system and it should explain every choice of them (such as the thematic list, all issues, sub-issues, indicators, data, computational methods, external sources, etc.). The content of the "metadata box" should be written in such a way, so that at a later stage, experts, belonging to the same scientific discipline, can understand any special characteristic and any limitation of the use of all entities in the decision support system. The level of description must be detailed enough, so that these other experts can take over the maintenance and upgrade of the DS system. We must emphasize that confidentiality and/or intellectual property issues can be resolved by letting:

(a) The external sources control access to sensitive information, that is, information covered by GDPR; and/or, probably,
(b) The administration of the main core of the DSS have analogous rights.

Preferably, all this information must be further organized by building a specifically oriented sub-meta-database, which will provide all possible ways for accessing this information (e.g., accredited keywords, characteristic terms and methodology names). Additionally, a knowledge base [50,51] may be included in the DS system, with references to other related, important information.

5.5. Incorporating the Proper Documentation and Navigation Information for the Users in the Teleological Meta-Database

As it has been many times emphasized before, a large-scale decision support system is very complex. Thus, even a user expert in the related scientific field cannot use the system properly unless he/she is suitably advised. Therefore, it is rather imperative that specific documentation and navigation information must be included in the meta-database, so that a user, who is not familiar with the DS system, can benefit from it with minimal difficulty. There are two aspects of the aforementioned kind of information:

- The first is scientific: The expert designer of the system has to provide proper content titles or even content synopses, that can facilitate the unfamiliar user in navigating the DS system easily and without mistakes.
- The second one is technical: It aims at helping the user to understand the information system and the user interface. It must be provided by the ITC expert(s) who built the IS.

Overall, the documentation template of the meta-database can be built in such a manner, so as to assist all user experts and their collaborators to find out themselves how to use the DS system as easily as possible.

6. Substantial Advantages of a Decision Support System Developed on the Basis of the Proposed New Teleological Meta-Database

As it has been already discussed, the meta-database must have the following characteristics and properties:

(a) It must include the purpose and use of the employed indicators and/or of each datum.
(b) It must incorporate the entire thematic list, that is, all issues and sub-issues and the exact way each one of them is linked to the entities of (a) immediately above.
(c) All software programs of the DSS must be directly linked to the entities referred to in (a) and (b) immediately above.

(d) The aforementioned software programs of the DSS must be very well documented, and this documentation must be properly placed in the template of the meta-database in a clear-cut manner.

Consequently, the proposed metadata structure can be considered as an exact and thorough content description and functional map of the DS system in hand.

One main novelty of the methodology described here is that during the entire life cycle of the DS system, a group of experts and engineers, even radically different than the one that developed the specific information system, may achieve the following, by employing the proposed teleological meta-database:

(a) Correct and efficient updating of the DS system.
(b) Make changes to the content of the DSS.
(c) Upgrade and/or expand the content of the DS system, by including further classes of issues, sub-issues, data, indicators, associated computational methods, new external sources, together with the corresponding mediators, etc. In fact, by exploiting the structure of the meta-database, one may achieve a consistent, considerable growth of the DS system. Without the use of the proposed teleological meta-database, there is a serious risk that such growth of the DSS may render it unmanageable.
(d) Reduce the content of the DS system, by eliminating a subset of the aforementioned entities, without causing any damage to the remaining system at all.

We would like to emphasize that any change imposed on the system must be accurately reflected/incorporated in the meta-database template. In this way, the DSS will maintain all the very good properties mentioned above; in the opposite case, all aforementioned actions concerning the DS system will not be feasible.

We would like to point out that any part of the information of a DS system based on the meta-database may be georeferenced/geographical. Actually, the use of the meta-database permits a very advanced architecture of the georeferenced/geographical information, as it will be described in another work.

7. Conclusions

Recent advances in information technology and communications allow for the development of very large-scale decision support systems. However, the complexity, the polymorphism and the size of the data and/or program core of such a very large DS system generate a number of very serious problems, which are outlined here. These problems have to do with:

(a) The efficient updating of the DS system;
(b) The flexible expansion of the DSS;
(c) The consistent reduction of the information system;
(d) The ability to modify the system;
(e) The upgradability of the DSS, etc.

Moreover, it is argued that any ad hoc/non-strictly methodological development of a very large-scale DS system will definitely lead to very serious intrinsic problems.

In the present manuscript, it was shown that only a consistent, fully methodological approach can resolve all the aforementioned problems and it can drastically minimize the associated risks and hazards. Such a methodological approach was explicitly introduced here. This methodology was associated with the development of a very detailed, consistent and rigorous meta-database, which fully describes the DS system and it includes:

(a) A complete thematic decomposition;
(b) A complete list of all indicators and/or data and their acquisition or evaluation method(s);

(c) A complete list of all the external sources of the system, together with the way to access them;
(d) An absolutely sufficient and structured description of the experts' knowledge concerning the whole system;
(e) The proper documentation and navigation information for users in this (teleological) meta-database.

Author Contributions: Conceptualization, E.K., C.P. and A.B.; Investigation, G.T., N.V.K. and A.B.; Project administration, E.K.; Resources, G.T., N.V.K. and E.M.; Supervision, E.K. and C.P.; Validation, E.K., C.P., G.T., N.V.K., A.B. and E.M.; Writing—original draft, E.K., C.P. and G.T.; Writing—review & editing, G.T., N.V.K., A.B., E.M. and A.R.M. All authors have read and agreed to the published version of the manuscript.

Funding: This research received no external funding.

Conflicts of Interest: The authors declare no conflict of interest.

References

1. Fagerholt, K. A computer-based decision support system for vessel fleet scheduling—Experience and future research. *Decis. Support Syst.* **2004**, *37*, 35–47. [CrossRef]
2. Fagerholt, K.; Johnsen, T.A.V.; Lindstad, H. Fleet deployment in liner shipping: A case study. *Marit. Policy Manag.* **2009**, *36*, 397–409. [CrossRef]
3. Clemente, M.; Fanti, M.P.; Iacobellis, G.; Nolich, M.; Ukovich, W. A Decision Support System for User-Based Vehicle Relocation in Car Sharing Systems. *IEEE Trans. Syst. Man Cybern. Syst.* **2017**, *48*, 1283–1296. [CrossRef]
4. Bolman, B.; Jak, R.G.; Van Hoof, L. Unravelling the myth—The use of Decisions Support Systems in marine management. *Mar. Policy* **2018**, *87*, 241–249. [CrossRef]
5. Gil, M.; Wróbel, K.; Montewka, J.; Goerlandt, F. A bibliometric analysis and systematic review of shipboard Decision Support Systems for accident prevention. *Saf. Sci.* **2020**, *128*, 104717. [CrossRef]
6. Pieri, G.; Cocco, M.; Salvetti, O. A Marine Information System for Environmental Monitoring: ARGO-MIS. *J. Mar. Sci. Eng.* **2018**, *6*, 15. [CrossRef]
7. Caris, A.; Macharis, C.; Janssens, G.K. Decision support in intermodal transport: A new research agenda. *Comput. Ind.* **2013**, *64*, 105–112. [CrossRef]
8. Qaiser, F.H.; Ahmed, K.; Sykora, M.; Choudhary, A.K.; Simpson, M. Decision support systems for sustainable logistics: A review and bibliometric analysis. *Ind. Manag. Data Syst.* **2017**, *117*, 1376–1388. [CrossRef]
9. Dotoli, M.; Epicoco, N.; Falagario, M.; Seatzu, C.; Turchiano, B. A Decision Support System for Optimizing Operations at Intermodal Railroad Terminals. *IEEE Trans. Syst. Man Cybern. Syst.* **2016**, *47*, 487–501. [CrossRef]
10. Zhan, J.; Ling, A.; An, N.; Li, L.; Sha, Y.; Li, X.; Liu, G. Building a Practical Ontology for Emergency Response Systems. In Proceedings of the IEEE Computer Society: Proceedings on, International Conference in Computer Science and Software Engineering CSSE '08 2008, Hubei, China, 12–14 December 2008; Volume 4, pp. 222–225. [CrossRef]
11. Comfort, L.K.; Sungu, Y.; Johnson, D.; Dunn, M. Complex Systems in Crisis: Anticipation and Resilience in Dynamic Environments. *J. Contingencies Crisis Manag.* **2001**, *9*, 144–158. [CrossRef]
12. Sakellariou, S.; Tampekis, S.; Samara, F.; Sfougaris, A.; Christopoulou, O. Review of state-of-the-art decision support systems (DSSs) for prevention and suppression of forest fires. *J. For. Res.* **2017**, *28*, 1107–1117. [CrossRef]
13. Rauner, M.S.; Niessner, H.; Odd, S.; Pope, A.; Neville, K.; O'Riordan, S.; Sasse, L.; Tomic, K. An advanced decision support system for European disaster management: The feature of the skills taxonomy. *Central Eur. J. Oper. Res.* **2018**, *26*, 485–530. [CrossRef]
14. McCown, R.L. Changing systems for supporting farmers' decisions: Problems, paradigms, and prospects. *Agric. Syst.* **2002**, *74*, 179–220. [CrossRef]
15. Salewicz, K.A.; Nakayama, M. Development of a web-based decision support system (DSS) for managing large international rivers. *Glob. Environ. Chang.* **2004**, *14*, 25–37. [CrossRef]
16. Ayed, M.B.; Ltifi, H.; Kolski, C.; Alimi, A.M. A user-centered approach for the design and implementation of KDD-based DSS: A case study in the healthcare domain. *Decis. Support Syst.* **2010**, *50*, 64–78. [CrossRef]

17. Sisk, G.M.; Miles, J.C.; Moore, C.J. Designer Centered Development of GA-Based DSS for Conceptual Design of Buildings. *J. Comput. Civ. Eng.* **2003**, *17*, 159–166. [CrossRef]
18. Tavasszy, L.A.; Smeenk, B.; Ruijgrok, C.J. A DSS for modelling logistic chains in freight transport policy analysis. *Int. Trans. Op. Res.* **1998**, *5*, 447–459. [CrossRef]
19. Ocalir-Akunal, E.V. Decision support systems in transport planning. *Procedia Eng.* **2016**, *161*, 1119–1126. [CrossRef]
20. Ballis, A. Implementing the European Transport information System. *Transp. Res. Rec. J. Transp. Res. Board* **2006**, *1957*, 23–31. [CrossRef]
21. Yatskiv, I.; Yurshevich, E. Data Actualization Using Regression Models in Decision Support System for Urban Transport Planning. In Proceedings of the 10th International Conference on Dependability and Complex Systems DepCoS-RELCOMEX 2015, Brunów, Poland, 29 June–3 July 2015; Volume 365. [CrossRef]
22. Damart, S.; Roy, B. The uses of cost–benefit analysis in public transportation decision-making in France. *Transp. Policy* **2009**, *16*, 200–212. [CrossRef]
23. Barfod, M.B.; Salling, K.B.; Leleur, S. Composite decision support by combining cost-benefit and multi-criteria decision analysis. *Decis. Support Syst.* **2011**, *51*, 167–175. [CrossRef]
24. Sacchelli, S. A Decision Support System for trade-off analysis and dynamic evaluation of forest ecosystem services. *iFor.-Biogeosci. For.* **2018**, *11*, 171–180. [CrossRef]
25. Loseto, G.; Scioscia, F.; Ruta, M.; Gramegna, F.; Ieva, S.; Pinto, A.; Scioscia, C. Knowledge-Based Decision Support in Healthcare via Near Field Communication. *Sensors* **2020**, *20*, 4923. [CrossRef] [PubMed]
26. Spoladore, D.; Sacco, M. Semantic and Dweller-Based Decision Support System for the Reconfiguration of Domestic Environments: RecAAL. *Electronics* **2018**, *7*, 179. [CrossRef]
27. Moreira, M.W.L.; Rodrigues, J.J.P.C.; Korotaev, V.; Al-Muhtadi, J.; Kumar, N. A Comprehensive Review on Smart Decision Support Systems for Health Care. *IEEE Syst. J.* **2019**, *13*, 3536–3545. [CrossRef]
28. Encyclopaedia Britannica. Information System. Available online: https://www.britannica.com/topic/information-system (accessed on 6 December 2020).
29. Archive of the European Integration (AEI); European Sources Online (ESO). Information Guide: Trans-European Networks. University of Pittsburgh, Cardiff EDC, 2013. Available online: http://aei.pitt.edu/75428/ (accessed on 25 October 2020).
30. Port of Rotterdam. Port Community System. Available online: https://www.portofrotterdam.com/en/doing-business/services/service-range/port-community-system (accessed on 25 October 2020).
31. Gerard, J.; Gunn, H.; Walker, W. National and International Freight Transport Models: An Overview and Ideas for Future Development. *Transp. Rev.* **2004**, *24*, 103–124. [CrossRef]
32. De Jong, G.; Vierth, I.; Tavasszy, L.; Ben-Akiva, M. Recent developments in national and international freight transport models within Europe. *Transportation* **2013**, *40*, 347–371. [CrossRef]
33. Rich, J.; Nielsen, O.A.; Brems, C.; Hansen, C.O. Overall design of the Danish national transport model. In *Proceedings of the Annual Transport Conference*; Aalborg University: Aalborg, Denmark, 2010; Volume 17, ISSN 1603-9696.
34. Chiehyeon, L.; Kim, K.J.; Maglio, P. Smart Cities with Big Data: Reference models, challenges, and considerations. *Sci. Direct* **2018**, *82*, 86–99. [CrossRef]
35. Gerike, R.; Schulz, A. Workshop Synthesis: Surveys on long-distance travel and other rare events. *Transp. Res. Procedia* **2018**, *32*, 535–541. [CrossRef]
36. Sillitto, H.G. Design principles for Ultra-Large-Scale (ULS) Systems. *INCOSE Int. Symp.* **2010**, *20*, 63–82. [CrossRef]
37. Gabriel, R.P.; Northrop, L.; Schmidt, D.C.; Sullivan, K. Ultra-large-scale systems. In Proceedings of the Companion to the 21st ACM SIGPLAN Symposium on Object-Oriented Programming Systems, Languages, and Applications (OOPSLA 2006), Portland, OR, USA, 22–26 October 2006; Association for Computing Machinery: New York, NY, USA, 2006; pp. 632–634. [CrossRef]
38. Arnott, D.; Dodson, G. *Decision Support Systems Failure*; Springer: Berlin/Heidelberg, Germany, 2008; Volume 1, pp. 763–790. [CrossRef]
39. Jovic, M.; Tijan, E.; Zgaljic, D.; Karanikic, P. SWOT analysis of selected digital technologies in transport economics. In Proceedings of the 43rd International Convention on Information, Communication and Electronic Technology, Digitalizing Logistics processes (DIGLOGS), Opatija, Croatia, 28 September–2 October 2020.

40. Limitations & Disadvantages of Decision Support Systems. Available online: https://www.managementstudyguide.com/limitations-and-disadvantages-of-decision-support-systems.htm (accessed on 25 October 2020).
41. Disadvantages of Decision Support Systems. Available online: http://dsssystem.blogspot.com/2010/01/disadvantages-of-decision-support.html (accessed on 25 October 2020).
42. United Nations, Treaty Collection, Depositary, Environment. A Kyoto Protocol to the United Nations Framework Convention on Climate Change. Available online: https://treaties.un.org/pages/ViewDetails.aspx?src=TREATY&mtdsg_no=XXVII-7a&chapter=27&lang=en (accessed on 25 October 2020).
43. United Nations, Climate Change, Process and Meetings. The Kyoto Protocol-Status of Ratification. Available online: https://unfccc.int/process/the-kyoto-protocol/status-of-ratification (accessed on 25 October 2020).
44. Grubb, M.; Vrolijk, C.; Brack, D. *Routledge Revivals: Kyoto Protocol (1999): A Guide and Assessment*; Routledge: Abingdon-on-Thames, UK, 2018; ISBN 9781315147024. [CrossRef]
45. Oberthür, S.; Hermann, E.O. *The Kyoto Protocol: International Climate Policy for the 21st Century*; Springer Science & Business Media: Berlin, Germany, 1999. [CrossRef]
46. Dearle, A. *Software Deployment, Past, Present and Future*; Future of Software Engineering (FOSE): Minneapolis, MN, USA, 2007; pp. 269–284. [CrossRef]
47. Kavvadia, H. Teleology in Systems Analysis and Design Methods. *SSRN Electron. J.* **2020**. [CrossRef]
48. Méline, J.; Van Hulst, A.; Thomas, F.; Chaix, B. Road, rail, and air transportation noise in residential and workplace neighborhoods and blood pressure (RECORD Study). *Noise Health* **2015**, *17*, 308–319. [CrossRef] [PubMed]
49. Chen, L.; Yang, H. Managing congestion and emissions in road networks with tolls and rebates. *Transp. Res. Part B Methodol.* **2012**, *46*, 933–948. [CrossRef]
50. Atymtayeva, L.; Kozhakhmet, K.; Bortsova, G. Building a knowledge base for expert system in information security. *Soft Comput. Artif. Intell. Adv. Intell. Syst. Comput.* **2014**, *270*, 57–76. [CrossRef]
51. Breuil, D. On knowledge based management systems: Integrating artificial intelligence and database technologies. *Eur. J. Oper. Res.* **1988**, *33*, 354–355. [CrossRef]

Article

Efficient Rule Generation for Associative Classification

Chartwut Thanajiranthorn * and Panida Songram

Department of Computer Science, Faculty of Informatics, Mahasarakham University, Mahasarakham 44150, Thailand; panida.s@msu.ac.th

* Correspondence: chartwut@bru.ac.th; Tel.: +66-619-395-455

Received: 2 September 2020; Accepted: 14 November 2020; Published: 17 November 2020

Abstract: Associative classification (AC) is a mining technique that integrates classification and association rule mining to perform classification on unseen data instances. AC is one of the effective classification techniques that applies the generated rules to perform classification. In particular, the number of frequent ruleitems generated by AC is inherently designated by the degree of certain minimum supports. A low minimum support can potentially generate a large set of ruleitems. This can be one of the major drawbacks of AC when some of the ruleitems are not used in the classification stage, and thus (to reduce the rule-mapping time), they are required to be removed from the set. This pruning process can be a computational burden and massively consumes memory resources. In this paper, a new AC algorithm is proposed to directly discover a compact number of efficient rules for classification without the pruning process. A vertical data representation technique is implemented to avoid redundant rule generation and to reduce time used in the mining process. The experimental results show that the proposed algorithm archives in terms of accuracy a number of generated ruleitems, classifier building time, and memory consumption, especially when compared to the well-known algorithms, Classification-based Association (CBA), Classification based on Multiple Association Rules (CMAR), and Fast Associative Classification Algorithm (FACA).

Keywords: associative classification; class association rule; vertical data representation; classification

---

1. Introduction

Nowadays, there a number of classification techniques that have been applied to various real-world applications, i.e., graph convolutional networks for text classification [1], automated classification of epileptic electroencephalogram (EEG) signals [2], iris cognition [3], and anomaly detection [4]. Associative classification (AC) is a well-known classification technique that was first introduced by Lui et al. [5]. It is a combination of two data-mining techniques, association rule mining, and classification. Association rule mining discovers the relationship between items in a dataset. Meanwhile, classification aims to predict the class label of any given instance from learning-labeled dataset. AC focuses on finding Class Association Rules (CARs) that satisfy certain minimum support and confidence thresholds in the form $x \rightarrow c$, where x is a set of attribute values and c is a class label. AC has been reported in the literature to outperform other traditional classifiers [6–13], In addition, a CAR is an if–then rule that can be easily understood by general users. Therefore, AC is applied in many fields, i.e., phishing website detection [6,7,11], heart disease prediction [8,9], groundwater detection [12], and detection of low-quality information in social networks [10].

In traditional AC algorithms, minimum support threshold is a significant key parameter that is used to select frequent ruleitems and to then eliminate frequent ruleitems in which confidence values do not satisfy minimum confidence. This manner leads to a large number of frequent ruleitems. Nguyen and Nguyen [14] demonstrated that the number of 4 million frequent ruleitems can be generated when the minimum support threshold is set to 1%. Moreover, a number of AC-based techniques, i.e., Classification-based Association (CBA) [5], Fast Associative Classification Algorithm (FACA) [11], CAR-Miner-diff [14], Predictability-Based Class Collative Class Association Rules (PCAR) [15], Weighted Classification Based on Association Rules (WCBA) [16], and Fast Classification Based on Association Rules (FCBA) [17], create all possible CARs in order to determine a set of valid CARs that can be used in the classification process. Recently, Active Pruning Rules (APR) [13] has been proposed as a novel evaluation method. APR can be used to avoid generating all CARs. However, the exhaustive search for finding rules in classifiers may cause an issue in large datasets or low minimum support. Creating candidate CARs consumes intensive computational times and memory. The minimal process of candidate generation is still challenging because it is quite affected in terms of training time, input/output (I/O) overheads, and memory usage [18].

In this paper, a new algorithm is proposed to directly generate a small number of efficient CARs for classification. A vertical data format [19] is used to represent ruleitems associated with their transaction IDs. The intersection technique is used to easily calculate support and confidence values from the format. The ruleitems with 100% of confidence will be added to the classifier as a CAR. Whenever a CAR with 100% confidence is found, the transaction associated with the CAR will be removed by using a set difference to avoid generating redundant CARs. Finally, a compact classifier is built for classification. In conclusion, the contribution of this paper is as follows.

1. To avoid pruning and sorting processes, the proposed algorithm directly generates CARs with 100% confidence to build compact classifiers. The CARs with 100% confidence are anticipated to result in high prediction rates.
2. The proposed algorithm eliminates unnecessary transactions to avoid generating redundant CARs in each stage.
3. Simple set theories, intersection, and set difference are exploited to reduce computational time used in mining process and to reduce memory consumption.

This paper is structured as follows. In Section 2, related works of AC are described. The basic definitions are delineated in Section 3. The proposed algorithm is introduced in Section 4. The discussion on the experimental results is in Section 5. Lastly, the conclusion of the study is stated in Section 6.

2. Related Work

In the past, AC-based algorithms have been proposed and studied. The study's objective is to understand some drawbacks and to increase the effectiveness of the algorithms. Lui et al. [5] introduced the CBA algorithm which integrated association rule mining and classification. The process of the CBA algorithm is divided into two steps. First, CARs are generated based on the famous search method the Apriori algorithm [20]. Second, CARs are sorted and then pruned to select efficient CARs in a classifier. The CBA algorithm was proven to produce a lower error rate than C4.5 [21]. Unfortunately, the CBA algorithm encounters a large number of candidate generation problems due to Apriori inheritance which finds all possible frequent rules at each level.

Li et al. [22] presented the Classification based on Multiple Association Rules (CMAR) algorithm. Unlike CBA, CMAR adopts a Frequent pattern tree (FP-tree) and a Cosine R-tree (CR-tree) for rule generation and classification phases. It divides the subset in FP-tree to search frequent ruleitems and then adds the frequent ruleitems to CR-tree according to their frequencies. Hence, CMAR only needs to scan the database once. The CMAR algorithm uses multiple rules to predict unseen instances based on chi-square method. In the experiment, CMAR was compared with CBA and C4.5 in terms of accuracy. The experimental result shows that CMAR performs better than the others.

Abdelhamid [6] proposed an Enhanced Multi-label Classifier-based Associative Classification (eMCAC) for phishing website detection. It generates rules with multiple class labels from a single dataset without recursive learning. The eMCAC algorithm applies a vertical data format to represent datasets. The support and confidence values for a multi-label rule are calculated based on the average support and confidence values of all classes. The class is assigned to the test instance if attribute values are fully matched to the rule's antecedent. The experimental results show that the eMCAC algorithm outperforms CBA, PART, C4.5, jRiP, and MCAR [23] on the real-world phishing data in terms of accuracy.

Hadi et al. [11] proposed the FACA algorithm for phishing website detection. It applies a Diffset [24] in the rule-generation process to increase the speed of classifier building time. First, the FACA algorithm discovers k-ruleitems by extending frequent (k − 1)-ruleitems. Then, ruleitems are ranked according to the number of attribute values, confidence, support, and occurrence. To predict unseen data, the FACA algorithm utilizes the All Exact Match Prediction Method. The method matches unseen data with all CARs in the classifiers. Next, unseen data are assigned to the class label with the highest count. From the experimental result, the FACA algorithm outperforms CBA, CMAR, MCAR, and ECAR [25] in terms of accuracy.

Song and Lee [15] introduced Predictability-Based Collective Class Association Rule algorithm (PCAR) to enhance rule evaluation. The PCAR algorithm uses inner cross-validation between the test dataset and train dataset to calculate a predictability value of CARs. Then, CARs are ranked according to rule predictive values, rule confidence, rule support, rule antecedent length, and rule occurrences. Finally, the full-matching method is applied to assign a class label for unseen data. To evaluate the performance of PCAR, PCAR was compared with C4.5, RIPPER, CBA, and MCAR on the accuracy, and PCAR was shown to outperform the others.

Alwidian et al. [16] proposed the WCBA algorithm to enhance the accuracy of a classifier based on the weighting technique. WCBA assumes that the importance of attributes is not equal. For example, in medicine, some attributes are more important than other attributes for prediction. Consequently, weights of all attributes are assigned by experts in the domain. Then, the weighted method is used to select useful CARs and a statistical measure is used for the pruning process. In addition, CARs are priors sorted by using the harmonic mean, which is an average value between support and confidence. The WCBA algorithm is more significantly accurate than CBA, CMAR, MCAR, FACA, and ECBA. However, the WCBA algorithm generates CARs based on the Apriori technique that scans the database many times.

Rajab proposed [13] the Active Pruning Rule (APR) algorithm. The new pruning process was introduced in APR. CARs are ranked by confidence, support, and rule length. Each training instance is matched over a set of CARs. The first rule that matches an instance is added to the classifier. Then, instances containing the first rule are removed. The support and confidence of remaining rules are recalculated, and all CARs are re-ranked. The APR algorithm was proven to reduce the size of the classifier and to maintain predictive accuracy performance. However, the APR algorithm still has to face a massive number of candidates from a rule-generation process. From previous works, the advantages and disadvantages are shown in Table 1.

The previous algorithms on AC generally result in high predictability of rules. However, most of them produce k-ruleitems from (k − 1)-ruleitems. They have to calculate supports when a new ruleitems is recovered. To calculate support and confidence values, they have to search all transactions in databases multiple times. Moreover, a huge number of candidate CARs are generated and pruned later to reduce unnecessary CARs. To reduce the problems, the proposed algorithm will directly generate efficient CARs for classification so that the pruning and sorting processes are not necessary. The efficient CARs in our works are rules with 100% confidence which are generated based on the idea that some attribute values can immediately indicate the class label if all attribute values belong to a class label. To easily check attribute values belonging to any class label, vertical data representation is used in the proposed algorithm. Furthermore, simple set theories, intersection, and set difference are adapted to easily calculate support and confidence values without scanning a database multiple times.

Table 1. Advantages and disadvantages of Associative classification (AC) algorithms.

Algorithms	Advantage	Disadvantage
CBA	It adopted the association rule technique to classify data that is proven to be more accurate than the traditional classification technique.	It has to face a sensitivity of the minimum support threshold. A massive number of rules are generated when a low minimum support threshold is given.
CMAR	It uses an efficient FP-tree, which consumes less memory and space compared to CBA.	The FP-tree will not always fit in the main memory, especially when the number of attributes is large.
eMCAC	It adopts vertical data representation to reduce space usage to find a multi-label class.	It is based on an Apriori-like technique that can result in a large number of frequent itemsets.
FACA	Set difference is adopted to consume low memory and to reduce the mining time.	It is based on an Apriori-like technique; therefore, the algorithm is required to search for all frequent itemsets from all possible candidate itemsets at each level.
PCAR	It uses predictability value to prune unnecessary rules.	The execution time is slow since it includes the inner cross-validation phase for calculation predictability value.
WCBA	It uses a weighted method to select useful rules and to improve the performance of the classifier.	Weighted factors are subject to change due to the decisions of experts which can cause a different experimental result.
APR	A new evaluation method with a small classifier and high accuracy rate.	Generation of a large number of rules when a low minimum support threshold is given.

3. Basic Definitions

Let $A = \{a_1, a_2, \ldots, a_m\}$ be a finite set of all attributes in dataset. $C = \{c_1, c_2, \ldots, c_n\}$ is a set of classes, $g(x)$ is a set of transactions containing itemset x, and $|g(x)|$ is the number of transactions containing x.

Definition 1. *An item can be described as an attribute a_i containing a value v_j, denoted as (a_i, v_j).*

Definition 2. *An itemset is the set of items, denoted as $(a_{i1}, v_{i1}), (a_{i2}, v_{i2}), \ldots, (a_{ik}, v_{ik})$.*

Definition 3. *A ruleitem is of the form $\langle itemset, c_j \rangle$, which represents an association between itemsets and class in a dataset; basically, it is represented in the form itemset $\rightarrow c_j$.*

Definition 4. *The length of a ruleitem is the number of items, denoted as $k - ruleitem$.*

Definition 5. *The absolute support of ruleitem r is the number of transactions containing r, denoted as $sup(r)$. The support of r can be found from (1).*

$$sup(r) = |g(r)| \tag{1}$$

Definition 6. *The confidence of ruleitem $\langle itemset, c_j \rangle$ is the ratio of the number of transactions that contains the itemset in class in c_j and the number of transactions containing the itemset, as in (2).*

$$conf(\langle itemset, c_j \rangle) = \frac{|g(\langle itemset, c_j \rangle)|}{|g(itemset)|} \times 100 \tag{2}$$

Definition 7. *Frequent ruleitem is a ruleitem in which support is not less than the minimum support threshold ($minsup$).*

Definition 8. *Class Association Rule (CAR) is a frequent ruleitem in which confidence is not less than the minimum confidence threshold ($minconf$).*

4. The Proposed Algorithm

In this section, a new algorithm, called the Efficient Class Association Rule Generation (ECARG) algorithm, is presented. The pseudo code of the proposed algorithm is shown in Algorithm 1.

Algorithm 1: Efficient Class Association Rule Generation (ECARG) algorithm main process

```
   Input: dataset, minsup
   Output: classifier
1  ruleItems = 1-ruleitem generation from dataset
2  while at least one rule's support meet minsup do
3      R = maximum confidence rule from ruleItems // ruleitem's support ≥ minsup
4      if R's confidence < 100 and R is not null then
5          R = extend R with the other ruleitems
6      if R is not null then
7          insert R to classifier
8          redundant rule removal
9          update support and confidence for each ruleItems
10     else
11         exit while loop
12 finding the default class
13 return classifier
```

First, 1-frequent ruleitems are generated (line 1). To quickly find 1-frequent ruleitems, the proposed algorithm takes the advantage of a vertical data format to calculate the support of the ruleitems. The support of the ruleitems can be obtained from $|g(itemset) \cap g(c_k)|$. If any 1-ruleitem does not meet the minimum support threshold, it will not be extended with the other ruleitems. Moreover, the confidence of the frequent ruleitems can be calculated from Equation (2) by using the vertical data format. If the confidence of the ruleitem is 100%, the ruleitems will be added to the classifier directly (line 7); otherwise, it will be considered extended with the others (line 5).

After discovering the most effective CAR with 100% confidence, the transaction IDs associated with the CAR will be removed to avoid redundant CARs (line 8). To remove the transaction IDs, a set difference plays an important role in our algorithm. Let r_i be a CAR with 100% confidence and T be a set of ruleitems in the same class of r_i. For all $r_j \in T$, the new transaction IDs of r_j is $g(r_j) = g(r_j) - g(r_i)$. Then, the new transaction IDs, support, and confidence values of all rules are updated (line 9).

In each iteration, if there is no CAR with 100% confidence, the ruleitem r with the highest confidence will be first to be considered extended in a breadth-first search manner. It will be combined with other ruleitems in the same class until the new CAR has 100% confidence (line 5). If r_i is extended with r_j to be r_{new} and $g(r_j) \subseteq g(r_i)$, then $conf(r_{new}) = 100\%$. After the extended CAR is added to the classifier, the transaction IDs associated with the CAR will be removed. Finally, if no ruleitem satisfies the minimum support threshold, the CAR generation will be stopped.

The proposed algorithm continues to find a default class in order to insert it to the classifier. The class with the most remaining transaction IDs is selected as the default class (line 12).

To demonstrate the examples, the dataset in Table 2 is used as example data. The minimum support and confidence thresholds are set to 2 and 50%, respectively.

Table 2. A sample dataset.

TID	atr1	atr2	atr3	Class Label
1	a_1	b_1	c_1	A
2	a_1	b_1	c_2	A
3	a_1	b_2	c_1	A
4	a_1	b_3	c_1	A
5	a_2	b_1	c_2	B
6	a_2	b_2	c_2	B
7	a_2	b_3	c_1	B
8	a_3	b_2	c_2	A
9	a_2	b_3	c_1	A

The vertical data format represents associated transaction IDs of 1-ruleitem, as shown in Table 3. The last 2 columns of Table 3 show the support and confidence of ruleitems that are calculated. From Table 2, the a_2 value in $atr1$ occurs in transaction IDs 5, 6, 7, and 9, denoted as $g(\langle atr1, a_2\rangle) = \{5,6,7,9\}$. Class A is in transaction IDs 1, 2, 3, 4, 8, and 9, denoted as $g(A) = \{1,2,3,4,8,9\}$, while class B is in transaction IDs 5, 6, and 7, denoted as $g(B) = \{5,6,7\}$. The transaction IDs containing $\langle atr1, a_2\rangle \rightarrow A$ are $g(\langle atr1, a_2\rangle) \cap g(A) = \{5,6,7,9\} \cap \{1,2,3,4,8,9\} = \{9\}$, so the supports of $\langle atr1, a_2\rangle \rightarrow A$ are 1. The rule $\langle atr1, a_2\rangle \rightarrow A$ will not be extended because its support is less than the minimum support threshold. Transaction IDs containing $\langle atr1, a_2\rangle \rightarrow B$ are $g(\langle atr1, a_2\rangle) \cap g(B) = \{5,6,7,9\} \cap \{5,6,7\} = \{5,6,7\}$, so the supports of $\langle atr1, a_2\rangle \rightarrow B$ are 3. Hence, this rule is a frequent ruleitem.

The confidence of $\langle atr1, a_2\rangle \rightarrow B$ can be obtained from $\frac{|g(5,6,7)|}{|g(5,6,7,9)|} \times 100 = \frac{3}{4} \times 100 = 75\%$. The confidence of $\langle atr1, a_2\rangle \rightarrow B$ is not 100% so it will be extended, whereas the confidence of $\langle atr1, a_1\rangle \rightarrow A$ is $\frac{|g(1,2,3,4)|}{|g(1,2,3,4)|} \times 100 = \frac{4}{4} \times 100 = 100\%$, so it is the first CAR added to the classifier.

Table 3. The rules that meet minimum support threshold (white background cell).

Ruleitem	TIDs	Sup	Conf (%)
$\langle atr1, a_1\rangle \to A$	$1, 2, 3, 4$	4	100
$\langle atr1, a_2\rangle \to A$	9	1	-
$\langle atr1, a_2\rangle \to B$	$5, 6, 7$	3	75
$\langle atr1, a_3\rangle \to A$	8	1	-
$\langle atr2, b_1\rangle \to A$	$1, 2$	2	66.67
$\langle atr2, b_1\rangle \to B$	5	1	-
$\langle atr2, b_2\rangle \to A$	3	1	-
$\langle atr2, b_2\rangle \to B$	6	1	-
$\langle atr2, b_3\rangle \to A$	$4, 8, 9$	3	75
$\langle atr2, b_3\rangle \to B$	7	1	-
$\langle atr3, c_1\rangle \to A$	$1, 3, 4, 9$	3	80
$\langle atr3, c_1\rangle \to B$	7	1	-
$\langle atr3, c_2\rangle \to A$	$2, 8$	2	50
$\langle atr3, c_2\rangle \to B$	$5, 6$	2	50

After discovering the first CAR, the transaction IDs associated with the CAR will be removed. From Table 3, if $\langle atr1, a_1\rangle$ is found, the class will absolutely be A. Hence, $\langle atr1, a_1\rangle \to A$ does not need to be extended with the other attribute values and transaction IDs 1, 2, 3, and 4 should be removed. The ECARG algorithm adopts a set difference, which can help to remove transaction IDs more conveniently.

For example, $g(\langle atr1, a_1\rangle \to A) = \{1,2,3,4\}$ and $g(\langle atr3, c_1\rangle \to A) = \{1,3,4,9\}$. The new transaction IDs of $g(\langle atr3, c_1\rangle \to A) = g(\langle atr3, c_1\rangle \to A) - g(\langle atr1, a_1\rangle \to A) = \{1,3,4,9\} - \{1,2,3,4\} = \{9\}$. Then, the new transaction IDs, support, and confidence values of all rules are updated as shown in Table 4.

From Table 4, there is no CAR with 100% confidence. $\langle atr1, a_2\rangle \to B$ has the maximum confidence, and $\langle atr3, c_2\rangle \to B = \{5,6\}$ is a subset of $g(\langle atr1,\ a_2\rangle \to B) = \{5,6,7\}$. Hence, the new rule $\langle (atr1, a_2), (atr3, c_2)\rangle \to B$ is found with 100% confidence. Then the extension of $\langle (atr1, a_2), (atr3, c_2)\rangle \to B$ is stopped. For 2-ruleitem extended from $\langle atr1, a_2\rangle \to B$, there is only one rule with 100% confidence and it is added to the classifier as the second CAR.

Table 4. The remained transaction IDs after generating the first Class Association Rule (CAR).

Ruleitem	TIDs	Sup	Conf (%)
$\langle atr1, a_2\rangle \to A$	9	1	-
$\langle atr1, a_2\rangle \to B$	$5, 6, 7$	3	75
$\langle atr1, a_3\rangle \to A$	8	1	-
$\langle atr2, b_1\rangle \to B$	5	1	-
$\langle atr2, b_2\rangle \to B$	6	1	-
$\langle atr2, b_3\rangle \to A$	$8, 9$	2	66.67
$\langle atr2, b_3\rangle \to B$	7	1	-
$\langle atr3, c_1\rangle \to A$	9	1	-
$\langle atr3, c_1\rangle \to B$	7	1	-
$\langle atr3, c_2\rangle \to A$	8	1	-
$\langle atr3, c_2\rangle \to B$	$5, 6$	2	66.67

After the second CAR is added to classifiers, the transaction IDs associated with CAR are removed. The remaining transaction IDs are shown in Table 5. There is only one ruleitem that satisfies the minimum support threshold: the ruleitem $\langle atr2, b_3 \rangle \rightarrow A$ which does not meet 100% of confidence. No ruleitem passes the minimum support threshold to be extended with the ruleitem $\langle atr2, b_3 \rangle \rightarrow A$ so CAR generation is stopped.

Table 5. Transaction IDs after generating the second CAR.

Ruleitem	TIDs	Sup	Conf (%)
$\langle atr1, a_2 \rangle \rightarrow A$	9	1	-
$\langle atr1, a_2 \rangle \rightarrow B$	7	1	-
$\langle atr1, a_3 \rangle \rightarrow A$	8	1	-
$\langle atr2, b_3 \rangle \rightarrow A$	8, 9	2	66.67
$\langle atr2, b_3 \rangle \rightarrow B$	7	1	-
$\langle atr3, c_1 \rangle \rightarrow A$	9	1	-
$\langle atr3, c_1 \rangle \rightarrow B$	7	1	-
$\langle atr3, c_2 \rangle \rightarrow A$	8	1	-

With the remaining transaction IDs in Table 5, the ECARG algorithm continues to find a default class and to add it to the classifier. In this step, the class with the most relevant transaction IDs is selected as the default class. In Table 5, class A remains in transaction IDs 8 and 9 while class B remains in transaction ID 7. The remaining transaction IDs are relevant to class A the most, so the default class is A. In case the number of associated remaining transaction IDs with each class is not changed, the majority class in the classifier is the default class. Finally, all CARs in the classifier are shown in Table 6.

Table 6. All CARs from ECARG.

CAR ID	CAR
R1	$\langle atr1, a_1 \rangle \rightarrow A$
R2	$\langle (atr1, a_2), (attr3, c_2) \rangle \rightarrow B$
Default Class	A

To observe the effect of 100% confidence ruleitems, we tested another version of ECARG, ECARG2. The difference in ECARG2 is ruleitem extension. If a ruleitem with 100% confidence cannot be found from the extension, the ruleitem with the highest confidence will be selected as a CAR and added to classifiers. For example, in Table 5, ruleitem $\langle atr2, b_3 \rangle \rightarrow A$ is the only ruleitem that satisfies the minimum support and minimum confidence. Hence, ECARG2 selects the ruleitem as the third CAR. The associated transaction IDs are removed, and the remaining transaction ID is shown in Table 7. There is only one transaction ID with class B. Consequently, the default class is B. Finally, all CARs from ECARG2 are shown in Table 8.

Table 7. Transaction IDs after ECARG2 generated the third CAR.

Rule Item	TIDs	Sup	Conf (%)
$\langle atr1, a_2 \rangle \rightarrow B$	7	1	-
$\langle atr2, b_3 \rangle \rightarrow B$	7	1	-
$\langle atr3, c_1 \rangle \rightarrow B$	7	1	-

Table 8. All CARs from ECARG2.

CAR ID	CAR
R1	$\langle atr1, a_1 \rangle \rightarrow A$
R2	$\langle (atr1, a_2), (attr3, c_2) \rangle \rightarrow B$
R3	$\langle atr2, b_3 \rangle \rightarrow A$
Default Class	B

5. Experimental Setting and Result

The experiments were implemented and tested on a system with the following environment: Intel Core i3-6100u 2.3 GHz processor with 8 GB DDR4 main memory, running Microsoft Windows 10 64-bit version. Our algorithm is compared with the well-known algorithms CBA, CMAR, and FACA. All algorithms were implemented in java. The implementing java version of the CBA algorithm using CR-tree is from WEKA [26]. The implementation of CMAR in JAVA is from [27]. Four algorithms are tested on 14 datasets from the UCI Machine Learning Repository. The characteristics of the datasets are shown in Table 9. Ten-fold cross-validation is used to divide testing instances and training instances based on previous works [12,17,23,26,27]. Accuracy rates, the number of CARs, classifier building times, and memory consumption are used to measure the performance of the four algorithms.

Table 9. Characteristics of the experiment datasets.

Data Sets	# of Attributes	# of Classes	Instances
Anneal	38	6	798
Breast	11	2	699
Cars	6	4	1,728
Contact-lenses	4	3	24
Diabetes	7	2	768
Iris	4	3	150
Labor	17	2	57
Lymph	18	4	148
Mushroom	22	2	8214
Post-operative	9	4	90
Tic-tac-toe	9	2	958
Vote	16	2	435
Wined	13	3	178
Zoo	17	7	101

To study the sensitivity of thresholds on the ECARG algorithm, we set different minimum support thresholds and different minimum confidence thresholds in the experiment. First, we set the minimum support thresholds from 1% to 4% and analyze different minimum confidence thresholds between 60%, 70%, 80%, and 90%. Figure 1 shows the accuracy rates of all datasets. The results show that, when the minimum support thresholds are increased, the accuracy rates are decreased. If the minimum confidence thresholds are increased, the accuracy rates are slightly down.

The highest accuracy rates are given in most datasets, Diabetes, Iris, Labor, Lymph, Mushroom, Post-operative, Tic-tac-toe, Vote, Wine, and Zoo, when minimum support and minimum confidence are set to 2% and 60%, respectively. Therefore, the minimum support is set to 2%, and minimum confidence is set to 60% in the next experiments.

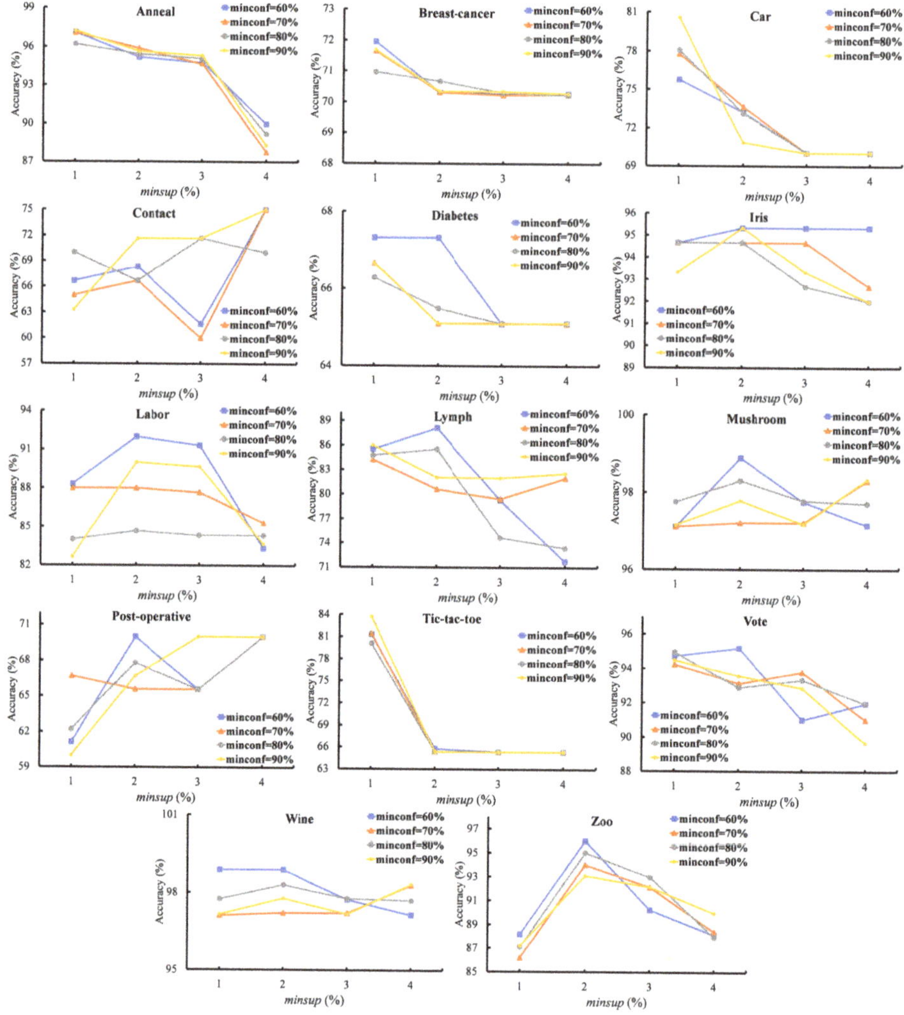

Figure 1. Accuracy rates in various *minsup* and *mincof* on all datasets.

Table 10 reports the accuracy rates of the CBA, CMAR, FACA, ECARG, and ECARG2 algorithms on the UCI datasets. The results show that both of our algorithms outperform the others on average. This gain resulting from the methodology found the most efficient rule in each iteration and eliminated redundant rules simultaneously. To be more precise, we further analyzed the win-lost-tie records. Based on Table 10, the win-lost-tie records of the ECARG2 algorithm against CBA, CMAR, FACA, and ECARG in terms of accuracy are 11-3-0, 11-3-0, and 9-4-1, 8-6-0, respectively. We can observe that ECARG gives an accuracy slightly less than ECARG2. However, the ECARG algorithm results in the highest accuracy in 6 of 14 datasets.

Table 10. Accuracies of CBA, CMAR, FACA, ECARG, and ECARG2.

Datasets	CBA	CMAR	FACA	ECARG	ECARG2
Anneal	83.19	73.27	87.31	95.21	**96.77**
Breast	67.16	**74.83**	72.44	70.33	73.02
Cars	78.29	73.73	70.02	73.43	**87.79**
Contact	66.67	37.5	63.33	**70.83**	65.00
Diabetes	**74.47**	57.03	73.56	67.32	73.7
Iris	92.67	**97.33**	96.00	95.33	96.00
Labor	75.67	26.32	87.67	**92.67**	84.00
Lymph	77.76	43.24	82.43	**88.51**	81.81
Mushroom	93.40	86.25	96.52	98.15	**98.9**
Post-oper.	56.67	**70.00**	67.78	**70.00**	60.00
Tic-tac-toe	**99.16**	53.03	90.23	65.34	88.94
Vote	94.02	92.64	91.92	**95.31**	95.17
Wine	89.97	62.92	92.16	**98.87**	97.16
Zoo	60.27	79.21	86.00	95.00	**96.00**
Average	79.24	66.24	82.67	84.02	**84.42**

Table 11 shows the average number of CARs generated from CBA, CMAR, FACA, ECARG, and ECARG2 algorithms. The result shows that the CMAR algorithm generates the highest number of rules, while the ECARG algorithm generates the lowest. In particular, the ECARG algorithm generates 8 CARs on average against 14 datasets whereas the CBA, CMAR, FACA, and ECARG2 algorithms derive 19, 240, 13, and 18 CARs on average, respectively. The accomplishment of the proposed algorithm is the discovery of the most efficient CAR in each iteration and the elimination of unnecessary transaction IDs that leads to redundant CARs.

Table 11. The average number of generated rules on the UCI datasets.

Data Sets	CBA	CMAR	FACA	ECARG	ECARG2
Anneal	**3**	165	15	14	17
Breast	16	127	23	**3**	35
Cars	25	272	9	**5**	18
Contact	9	25	**5**	7	8
Diabetes	56	115	24	**4**	38
Iris	11	38	7	**4**	9
Labor	**8**	297	15	9	9
Lymph	26	465	**15**	19	20
Mushroom	**8**	28	16	12	13
Post-oper.	35	51	12	**11**	27
Tic-tac-toe	28	713	12	**6**	33
Vote	30	658	12	11	**10**
Wined	**5**	237	11	9	7
Zoo	**10**	97	11	**10**	**10**
Average	19	240	13	**8**	18

Table 12 shows the average classifier building time of the proposed algorithm against CBA, CMAR, and FACA. The experimental result clearly shows that our algorithm is the fastest among all algorithms in the 14 datasets. ECARG takes fewer seconds to construct the classifier than CBA, CMAR, FACA, and ECARG2 by 2.134, 0.307, 2.883, and 0.0162, respectively. This can be explained by the fact that CBA and FACA uses an Apriori-style approach to generate candidates. When the value for minimum support is low on large datasets, it is costly to handle a large number of candidate ruleitems. The CMAR algorithm based on FP-growth is better than CBA and FACA in some cases, but it takes more classifier-generating time than ECARG and ECARG2.

Table 12. The classifier building time in seconds.

Data Sets	CBA	CMAR	FACA	ECARG	ECARG2
Anneal	1.050	**0.098**	0.877	0.123	0.164
Breast	0.670	0.169	0.185	**0.007**	0.027
Cars	0.220	0.249	0.640	**0.057**	0.062
Contact	0.010	0.075	0.004	**0.001**	0.002
Diabetes	1.160	0.107	0.558	**0.032**	0.085
Iris	0.030	0.008	0.010	**0.004**	**0.004**
Labor	1.170	0.924	0.027	**0.005**	**0.005**
Lymph	1.320	3.782	3.700	**0.016**	**0.016**
Mushroom	25.830	**0.104**	21.500	4.049	4.128
Post-oper.	0.090	0.041	0.063	**0.008**	0.012
Tic-tac-toe	0.230	0.235	0.800	**0.101**	0.135
Vote	1.540	2.601	5.300	**0.034**	0.034
Wined	0.120	0.273	0.190	**0.007**	**0.007**
Zoo	0.900	0.005	0.047	0.020	**0.013**
Average	2.453	0.623	2.422	**0.319**	0.335

Table 13 reveals the memory consumption in the classifier building process of all 5 algorithms. The results show that ECARG consumes less memory than CBA, CMAR, FACA, and ECARG2 by 22.62 MB, 73.15 MB, 36.57 MB, and 0.98 MB, respectively. The memory consumption of ECARG is the best since it eliminates unnecessary data in each iteration. From the result in Table 14, our proposed algorithm gives a higher F-measure on average than the other algorithms. In particular, the ECARG2 outperformed CBA, CMAR, FACA, and ECARG by 3.82%, 25.38%, 25.38%, 12.74%, and 1.84%, respectively.

Table 13. The classifier building memory consumption in megabytes.

Data Sets	CBA	CMAR	FACA	ECARG	ECARG2
Anneal	73.47	29.16	10.78	**10.68**	13.38
Breast	25.44	23.96	24.4	**1.92**	3.54
Cars	60.08	21.17	8.98	**3.05**	3.76
Contact	2.65	**0.99**	1.87	1.78	1.84
Diabetes	28.08	26.74	24.61	**3.01**	7.30
Iris	4.16	2.40	1.88	**1.17**	**1.17**
Labor	18.34	420.88	124.01	**1.95**	**1.95**
Lymph	27.31	250.93	231.75	**2.86**	**2.86**
Mushroom	28.89	29.12	24.52	**24.27**	24.31
Post-oper.	15.17	8.78	16.38	**2.03**	2.61
Tic-tac-toe	31.76	62.23	12.76	**4.73**	8.44
Vote	23.57	2.65	3.13	**3.09**	3.15
Wine	20.87	175.36	59.52	**1.82**	**1.82**
Zoo	21.41	34.33	31.93	2.20	**2.13**
Average	27.23	77.76	41.18	**4.61**	5.59

Table 14. F-measure of Classification-based Association (CBA), Classification based on Multiple Association Rules (CMAR), Fast Associative Classification Algorithm (FACA), ECARG, and ECARG2.

Data Sets	CBA	CMAR	FACA	ECARG	ECARG2
Anneal	75.93	43.73	43.64	61.24	**89.28**
Breast	66.15	66.32	66.39	58.42	**68.43**
Cars	**73.85**	33.31	31.04	45.81	71.57
Contact	53.31	43.08	49.94	**71.67**	61.67
Diabetes	**74.4**	49.01	74.3	56.56	71.21
Iris	92.70	**97.98**	93.56	90.41	96.16
Labor	70.85	28.29	75.46	**88.89**	85.87
Lymph	78.83	48.14	53.63	**81.94**	72.08
Mushroom	93.75	87.61	96.52	96.58	**98.90**
Post-oper.	52.51	20.59	**56.00**	54.45	39.57
Tic-tac-toe	**98.90**	43.88	64.40	95.44	87.64
Vote	**94.95**	72.82	91.82	93.82	94.44
Wine	87.03	69.14	92.47	**98.65**	94.37
Zoo	54.61	62.00	53.73	**91.76**	89.37
Average	76.27	54.71	67.35	78.25	**80.09**

Table 15 shows standard deviations of accuracy rate, the number of generated rules, building times, memory consumption, and F-measure of ECARG. The standard deviation values of building time and memory consumption are low and show that the building time and memory consumption in each fold is approximately marginal. The standard deviation values of the number of generated rules are relevant.

The standard deviation values of accuracy rates and F-measure show that the values of accuracy rates and F-measure in each fold are marginally different on almost all datasets. However, when evaluating the small datasets, Contact-lenses, Labor, Lymph, and Post-operative, the standard deviation values are high because 10-fold cross-validation splits a very small testing set that can potentially affect the efficiency of the classifier. For example, the Contact-lenses dataset composes only 2 or 3 transactions in each testing set. Consequently, only one false classification occurs in the testing set and then reduces the accuracy rate dramatically.

Table 15. Standard deviations of ECARG.

Data Sets	Accuracy		# of Rules		Building Time		Memory		F-1	
	AVG	S.D.	AVG	S.D.	AVG	S.D.	AVG	S.D.	AVG	S.D.
Anneal	95.21	2.10	14	0.94	0.123	0.05	10.68	0.03	61.24	6.98
Breast	70.33	5.10	3	1.40	0.007	0.01	1.92	0.02	58.42	1.78
Car	73.43	6.11	5	0.82	0.057	0.05	3.05	0.00	45.81	7.16
Contact	70.83	28.81	7	0.92	0.001	0.00	1.78	0.02	71.67	30.54
Diabetes	67.32	6.75	4	1.34	0.032	0.01	3.01	0.03	56.56	3.41
Iris	95.33	5.44	4	0.52	0.004	0.00	1.17	0.00	90.41	5.48
Labor	92.67	14.05	9	1.26	0.005	0.00	1.95	0.02	88.89	13.72
Lymph	88.51	10.00	19	3.37	0.016	0.00	2.86	0.00	81.94	15.5
Mushroom	98.15	0.32	12	0.00	4.049	0.64	24.27	0.03	96.58	0.31
Post-oper	70.00	17.41	11	3.34	0.008	0.00	2.03	0.02	54.45	14.21
Tic-tac-toe	65.34	6.16	6	2.13	0.101	0.06	4.73	0.06	95.44	3.85
Vote	95.31	2.72	11	2.26	0.034	0.01	3.09	0.03	93.82	2.84
Wined	98.87	2.34	9	0.53	0.007	0.00	1.82	0.03	98.65	2.31
Zoo	95.00	6.99	10	0.70	0.020	0.01	2.20	0.03	91.76	13.65
Average	84.02	8.16	8	1.395	0.320	0.06	4.61	0.02	78.25	8.70

From the experimental results, the ECARG algorithm outperforms CBA, CMAR, and FACA in terms of accuracy rate and the number of generated rules. A key achievement of the ECARG algorithm is that the technique generates valid rules with 100% confidence to build classifiers. The high confidence demonstrates the high possibility of class occurrences occurring in an itemset. Therefore, the ECARG algorithm produces a small classifier but gives high accuracy. While the CBA, CMAR, and FACA algorithms build classifiers from CARs that meet the minimum confidence threshold, some of the CARs have low confidences so they may predict incorrect classese and then the accuracies of CBA, CMAR, and FACA are lower than the proposed algorithm in the most dataset.

Moreover, ECARG outperforms the others in terms of building time and memory consumption. This key achievement applies simple set theories, i.e., intersection and set difference, processing on vertical data, which can potentially reduce time and memory consumption. Furthermore, the search space can be reduced as unnecessary transactions are eliminated in each stage and, therefore, the classifier building time is minimized.

6. Conclusions

This paper proposes algorithms to enhanced associative classification. Unlike the traditional algorithms, the proposed algorithms do not need a sorting and pruning process. Candidate generation is carried out by attempting to select a first general rule with the highest accuracy. Moreover, a search space is reduced early by cutting down items with low statistical significance. Furthermore, a vertical data format, intersection, and set difference methods are applied to calculate support and confidence and to remove unnecessary transaction IDs, decreasing computation time and memory consumption.

The experiments were conducted on 14 UCI datasets. The experimental results show that the ECARG algorithm outperforms the CBA, CMAR, and FACA algorithms in terms of accuracy by 4.78%, 17.79%, and 1.35%, respectively. Furthermore, ECARG generates smaller rules than the other algorithms in almost all datasets. In addition, ECARG results in the most optimal classifier-generating time and memory usage on average. We can conclude that the proposed algorithm gives a compact classifier with a high accuracy rate, improves computation time, and reduces memory usage.

However, the ECARG algorithm does not well perform on imbalanced datasets, such as Breast, Car, Diabetes, and Post-operative. This is because the ECARG algorithm tends to find 100% confidence CARs and to eliminate unnecessary transactions. Therefore, ruleitems belonging to minority classes will not meet the minimum support threshold or 100% confidence and they are eliminated accordingly. Consequently, the classifier cannot classify the minority class correctly.

Author Contributions: Methodology, C.T.; supervision, P.S. All authors have read and agreed to the published version of the manuscript.

Funding: This research was financially supported by Mahasarakham University (Grant year 2021).

Conflicts of Interest: The authors declare no conflict of interest.

References

1. Yao, L.; Mao, C.; Luo, Y. Graph convolutional networks for text classification. In Proceedings of the AAAI Conference on Artificial Intelligence, Honolulu, HI, USA, 27 January–1 February 2019; Volume 33, pp. 7370–7377.
2. Jukic, S.; Saracevic, M.; Subasi, A.; Kevric, J. Comparison of Ensemble Machine Learning Methods for Automated Classification of Focal and Non-Focal Epileptic EEG Signals. *Mathematics* **2020**, *8*, 1481. [CrossRef]
3. Adamović, S.; Miškovic, V.; Maček, N.; Milosavljević, M.; Šarac, M.; Saračević, M.; Gnjatović, M. An efficient novel approach for iris recognition based on stylometric features and machine learning techniques. *Future Gener. Comput. Syst.* **2020**, *107*, 144–157. [CrossRef]
4. Ruff, L.; Vandermeulen, R.; Goernitz, N.; Deecke, L.; Siddiqui, S.A.; Binder, A.; Müller, E.; Kloft, M. Deep one-class classification. In Proceedings of the International Conference on Machine Learning, Stockholm, Sweden, 10–15 July 2018; pp. 4393–4402.
5. Liu, B.; Yiming, M.; Hsu, W. Integrating Classification and Association Rule Mining. In Proceedings of the Fourth International Conference on Knowledge Discovery and Data Mining, New York, NY, USA, 27–31 August 1998.
6. Abdelhamid, N. Multi-label rules for phishing classification. *Appl. Comput. Inform.* **2015**, *11*, 29–46. [CrossRef]
7. Abdelhamid, N.; Ayesh, A.; Thabtah, F. Phishing detection based associative classification data mining. *Expert Syst. Appl.* **2014**, *41*, 5948–5959. [CrossRef]
8. Jabbar, M.; Deekshatulu, B.; Chandra, P. Heart Disease Prediction System using Associative Classification and Genetic Algorithm. *arXiv* **2013**, arXiv: 1303.5919.
9. Singh, J.; Kamra, A.; Singh, H. Prediction of heart diseases using associative classification. In Proceedings of the 5th International Conference on Wireless Networks and Embedded Systems (WECON), Rajpura, India, 14–16 October 2016; pp. 1–7. [CrossRef]
10. Wang, D. Analysis and detection of low quality information in social networks. In Proceedings of the 2014 IEEE 30th International Conference on Data Engineering Workshops, Chicago, IL, USA, 31 March–4 April 2014; pp. 350–354. [CrossRef]
11. Hadi, W.; Aburub, F.; Alhawari, S. A new fast associative classification algorithm for detecting phishing websites. *Appl. Soft Comput.* **2016**, *48*, 729–734. [CrossRef]
12. Hadi, W.; Issa, G.; Ishtaiwi, A. ACPRISM: Associative classification based on PRISM algorithm. *Inf. Sci.* **2017**, *417*, 287–300. [CrossRef]

13. Rajab, K.D. New Associative Classification Method Based on Rule Pruning for Classification of Datasets. *IEEE Access* **2019**, *7*, 157783–157795. [CrossRef]
14. Nguyen, L.; Nguyen, N.T. An improved algorithm for mining class association rules using the difference of Obidsets. *Expert Syst. Appl.* **2015**, *42*, 4361–4369. [CrossRef]
15. Song, K.; Lee, K. Predictability-based collective class association rule mining. *Expert Syst. Appl.* **2017**, *79*, 1–7. [CrossRef]
16. Alwidian, J.; Hammo, B.H.; Obeid, N. WCBA: Weighted classification based on association rules algorithm for breast cancer disease. *Appl. Soft Comput.* **2018**, *62*, 536–549. [CrossRef]
17. Alwidian, J.; Hammo, B.; Obeid, N. FCBA: Fast Classification Based on Association Rules Algorithm. *Int. J. Comput. Sci. Netw. Secur.* **2016**, *16*, 117.
18. Abdelhamid, N.; Jabbar, A.A.; Thabtah, F. Associative classification common research challenges. In Proceedings of the 2016 45th International Conference on Parallel Processing Workshops (ICPPW), Philadelphia, PA, USA, 16–19 August 2016; pp. 432–437.
19. Ogihara, Z.P.; Zaki, M.; Parthasarathy, S.; Ogihara, M.; Li, W. New algorithms for fast discovery of association rules. In Proceedings of the 3rd International Conference on Knowledge Discovery and Data Mining, Newport Beach, CA, USA, 14–17 August 1997.
20. Agrawal, R.; Srikant, R. Fast algorithms for mining association rules. In Proceedings of the 20th International Conference Very Large Data Bases, VLDB, Santiago, Chile, 12–15 September 1994; Volume 1215, pp. 487–499.
21. Quinlan, J. *C4.5: Programs for Machine Learning*; Morgan Kaufmann Publisher, Inc.: Los Altos, CA, USA, 1993.
22. Li, W.; Han, J.; Pei, J. CMAR: Accurate and efficient classification based on multiple class-association rules. In Proceedings of the 2001 IEEE International Conference on Data Mining, San Jose, CA, USA, 29 November–2 December 2001; pp. 369–376.
23. Thabtah, F.; Cowling, P.; Peng, Y. MCAR: Multi-class classification based on association rule. In Proceedings of the 3rd ACS/IEEE International Conference on Computer Systems and Applications, Cairo, Egypt, 6 January 2005. [CrossRef]
24. Zaki, M.; Gouda, K. Fast Vertical Mining Using Diffsets. In Proceedings of the Ninth ACM SIGKDD International Conference on Knowledge Discovery and Data Mining, Washington, DC, USA, 24–27 August 2003; ACM: New York, NY, USA, 2003; pp. 326–335. [CrossRef]
25. Hadi, W. ECAR: A new enhanced class association rule. *Adv. Comput. Sci. Technol.* **2015**, *8*, 43–52.
26. Mutter, S. Class JCBA. 2013. Available online: https://github.com/bnjmn/weka (accessed on 30 September 2018).
27. Padillo, F.; Luna, J.M.; Ventura, S. LAC: Library for associative classification. *Knowl.-Based Syst.* **2019**, *193*, 105432. [CrossRef]

MDPI
St. Alban-Anlage 66
4052 Basel
Switzerland
Tel. +41 61 683 77 34
Fax +41 61 302 89 18
www.mdpi.com

Algorithms Editorial Office
E-mail: algorithms@mdpi.com
www.mdpi.com/journal/algorithms